D0602868

SOLAR ECLIPSES
and the
IONOSPHERE

SOLAR ECLIPSES
and the
IONOSPHERE

A NATO Advanced Studies Institute held in Lagonissi, Greece,
May 26-June 4, 1969

Edited by

Michael Anastassiades

Professor of Physical Electronics
Athens University

and

Director, Ionospheric Institute
National Observatory of Athens
Athens, Greece

⊕ PLENUM PRESS · NEW YORK–LONDON · 1970

Library of Congress Catalog Card Number 71-119056

SBN 306-30480-5

© 1970 Plenum Press, New York
A Division of Plenum Publishing Corporation
227 West 17th Street, New York, N.Y. 10011

United Kingdom edition published by Plenum Press, London
A Division of Plenum Publishing Company, Ltd.
Donington House, 30 Norfolk Street, London W.C.2, England

Printed in the United States of America

PREFACE

It is a long-standing tradition in this country for any event related to solar activity to enjoy special importance. Because we enjoy the advantage of prolonged sunny periods, we consider the incidence of some thirty cloudy days each year as a personal affront.

I cannot say that we have the faculty to produce solar eclipses in order to justify an Institute on this subject to be held in Athens, but I can say that the occurrence of a solar eclipse over Greece would present, for optical and solar astronomers, the highest possibility of excellent conditions for observation.

We had this opportunity during the May 20, 1966 annular solar eclipse over North Africa and southern Europe. At that time we had the pleasure to collaborate with a large number of research groups of various nationalities, who installed optical and radio instruments near or in Lagonissi, which lay along the central path, in order to follow the eclipse.

A first meeting was then organized in order to discuss methods of data handling and availability of results. After a preliminary discussion it was suggested that a later meeting should be organized when final results were available.

Owing to the fact that a large number of disciplines were involved, it was imperative to select one topic and promote our knowledge in this special field. The choice went to the study of the ionosphere during solar eclipses. This was justified because the last international conference on this subject had taken place 15 years ago in Cambridge, and since then other manifestations, while very important, were not so specific. The second reason was that if this meeting, following the suggestion of the participants of the preliminary meeting of 1966, could again be held in Athens, the host country would choose a subject of special interest to its research workers.

I cannot claim that our research team at the University of
Athens has a very long tradition in ionospheric investigations.
It is only nine years since we installed, at Scaramanga near Athens,
an ionosonde, and this is not enough time to create scientific
tradition. However the advantages due to our location here offer
an opportunity for specializing in the field of solar-terrestrial
relations.

The principal advantage, already mentioned, is our clear sky.
Using this advantage, we have created a very good service of optical observations of the sun, which is supplemented by collaboration
with the American Air Weather Service, which provides excellent
instrumentation.

This activity in the optical solar field of observations can
be combined with other geophysical measurements. Our obvious
experience based on the direct correlation between optical extra-
solar events and variations observed on ionospheric parameters
created the evidence that the ionosphere is a photocell sensitive
to any change of solar flux. It would be then of importance to
investigate the behavior of ionospheric layers during the May 20,
1966 solar eclipse when its path crossed the Greek area from south-
west to northeast by using our network of three ionosondes and
all other devices and new techniques at our disposal: radio-astronomical, riometer, satellite, and partial-reflection-technique
measurements were performed by our group, and results from all the
above measurements were compared with data obtained directly by
rockets launched by NASA.

For all the above reasons our group selected the topic "Solar
Eclipses and the Ionosphere" for an Advanced Studies Institute to
be held in Athens. The meeting was a very successful one, and as
Director of this Institute I am deeply indebted to all our outstanding lecturers and participants for their very valuable contributions. The emphasis which the discussions during the Athens
eclipse meeting placed on the contribution of the solar corona
to the ionospheric eclipse phenomena, on the chemistry of ionospheric layers, on aeronomical problems which still remain open,
on the dynamics of the F layers, etc., will be of particular importance for all investigators during the solar eclipse over
Mexico and the USA on March 7, 1970.

We are indebted to the Advisory Panel of the NATO Science
Committee on Summer Schools, who sponsored this Institute for a
second time. Following the NATO policy, the Institute gives an
opportunity to young scientists from countries under development
and from other developed countries to stimulate their research
with lectures given by outstanding scientists and by participating

in a free and frank discussion. A successful Advanced Studies
Institute must give certain time for relaxation. We hope that
Lagonissi proved well suited for this purpose. The Director of
the Institute hopes to have done his best to keep lectures to a
suitable length and to offer a social program to relax all participants after their studies and discussions.

Michael Anastassiades
Professor, Athens University
Director of the Ionospheric Institute,
National Observatory of Athens

Director, NATO Advanced Studies
Institute on Solar Eclipses and the
Ionosphere, 1966 and 1969

CONTENTS

CONCLUDING REMARKS

CONTRIBUTORS

Anastassiades, Michael, Electronics Laboratory, University of Athens, Athens, Greece

Bovlatsos, Demetrius, Electronics Laboratory, University of Athens, Athens, Greece

Bowhill, S. A., Aeronomy Laboratory, Department of Electrical Engineering, University of Illinois, Urbana, Illinois

Bulat, T., Jeofosik Kürsüsü, Istanbul Universitesi, Istanbul, Turkey

Croom, D. L., Radio and Space Research Station, Dittor Park, Slough, Bucks., England

Friedman, H., United States Naval Research Laboratory, Washington, D. C.

Hale, L. C., Ionosphere Research Laboratory, The Pennsylvania State University, University Park, Pennsylvania

Haubert, A., Groupe de Recherches Ionosphériques, France

Kane, J. A., Laboratory for Space Sciences, NASA Goddard Space Flight Center, Greenbelt, Maryland

Krimigis, S. M., The Johns Hopkins University Applied Physics Laboratory, Silver Spring, Maryland

King, J. W., Radio and Space Research Station, Dittor Park, Slough, Bucks., England

Matsoukas, Demetrius, Electronics Laboratory, University of Athens, Athens, Greece

Mitra, A. P., National Physical Laboratory, New Delhi-12, India

Ozdogan, I., Jeofosik Kürsüsü, Istanbul Universitesi, Istanbul, Turkey

Rishbeth, H., Radio and Space Research Station, Dittor Park, Slough, Bucks., England

Rycroft, M. J., Physics Department, University of Southampton, Southampton, England

Smith, P. A., Radio and Space Research Station, Dittor Park, Slough, Bucks., England

Straka, R. M., Radio Astronomy Branch, Ionospheric Physics Laboratory, Air Force Cambridge Research Laboratories, Bedford, Massachusetts

Thomas, J. O., Physics Department, Imperial College, University of London, London, England

Tsagakis, Emmanuel, Electronics Laboratory, University of Athens, Athens, Greece

Vassy, A., Physique de l'Atmosphere, Université de Paris, Paris, France

THEORETICAL ASPECTS

IONOSPHERIC EFFECTS IN SOLAR ECLIPSES[*]

S. A. Bowhill

Aeronomy Laboratory

University of Illinois

Urbana, Illinois

1. Introduction

It is 14 years since the last international symposium on solar eclipses was held in London in 1955 (Beynon and Brown 1956). This interval of 14 years has represented the advent of the space age, and the results of this considerable period of enhanced activity on research on the earth's environment provide most of the material for this paper. My task is made easier by the fact that a summary has recently been prepared (Rishbeth 1968) of ionospheric theory pertinent to solar eclipses, based on a paper presented at the Summer School on Ionosphere held at Athens in 1966.

Most of the information derived about the ionosphere from solar eclipses comes from a comparison of the behavior of the ionosphere during eclipse conditions with its behavior at the same solar elevation angle under full-sun conditions. In this survey, I shall discuss first those aspects of solar radiation which are important in solar eclipses, and then discuss effects in the E, F and D regions in that order; which is in fact the order of increasing complexity.

2. Solar radiation in an eclipse

As the disc of the sun is covered by the moon during a total solar eclipse, the solar flux in each wavelength region is progressively reduced, relative to its uneclipsed value. The ratio of the instantaneous solar flux at a given wavelength to its uneclipsed value is termed the "eclipse function". For visible radiation, it is unity at first and fourth contact, and zero at second and third contact for a total eclipse. For wavelengths in the ultraviolet or x-ray region, however, its behavior is quite different in two ways. First, regions of increased brightness on the solar disc may cause the eclipse function to change irregularly with solar obscuration; and second, radiation from outside the visible disc may give an eclipse function which does not become zero even when the whole visible disc is covered (in other words, an eclipse which is total for visible light may be annular for other wavelengths).

[*]The preparation of this paper was supported by the National Aeronautics and Space Administration under grant NGR 013.

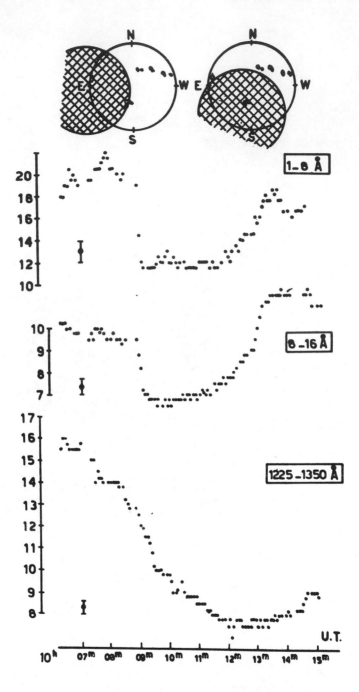

Figure 1.   Solar radiation in three bands measured by the SOLRAD 8 satellite
            (Landini et al., 1966).

As an example of a direct measurements of this first effect, Figure 1 (Landini et al., 1966) shows the variation of solar x-ray intensity in three wavelength regions during the solar eclipse of May 20, 1966, as seen by the SOLRAD 8 satellite. For ultraviolet radiation in the 1225-1350 A region, the eclipse function decreased fairly uniformly with time. At the x-ray wavelengths (1-8 and 8-16 A) which produce ionization in the D and lower E regions of the ionosphere, however, the solar disc was very non-uniform in brightness. Any analysis of D- and E-region ionization must use an eclipse function for the appropriate wavelength region, rather than simply the unobscured fraction of the visible disc. Unfortunately, it is quite unlikely that one of the few solar radiation satellites will be available in the path of any given eclipse, particularly during its maximum phase. Photographs of the solar disc using rocket spectrographs at various wavelengths are highly desirable, in that they show the contribution of active regions inside and outside the disc with greater resolution then can be obtained using available techniques from pictures of the full sun.

Our lack of knowledge as to the precise amount of coronal radiation which comes from outside the solar disc is particularly important for the interpretation of E-region ionospheric measurements during solar eclipses. For this region, x-ray and ultraviolet radiation are approximately equally important in producing ionization. Few observations are available of the residual x-ray intensity during totality. The Naval Research Laboratory group (Friedman 1962), using rocket-borne photometers during the solar eclipse of October 12, 1958, found a residual intensity at totality in the 44-60 Å x-ray band of about 10-13 percent of the full-sun emission. Smith et al. (1965) found 22 percent residual intensity during the partial eclipse of July 20, 1963 at Ft. Churchill, Canada, when the visible disc was 9.5 percent uncovered. Their best interpretation of the E-region ionization data was that half of the radiation followed the x-ray eclipse function, and half followed the unobscured area of the visible disc.

Lacking such detailed information as that illustrated in Figure 1, for most eclipses, it is difficult to make assumptions which are any more sophisticated than those just outlined. There is an obvious need for more accurate determinations of the x-ray intensity as a function of altitude in the solar atmosphere; these could be made, for example, by a rocket-borne spectrograph of relatively low angular resolution, fired during the early period of visible-light eclipse totality.

3. The E region during a solar eclipse

The typical behavior of E-region ionization during a solar eclipse is well illustrated by Figure 2, from Szendrei and McElhinny (1956). The minimum of electron density occures close to the time of second and third contact, and is about half that on the normal day. Also illustrated on Figure 2 are several curves found by solving the continuity equation for the ionization, on the assumptions that the rate of electron production was proportional to the uneclipsed area of the visible disc; and that the recombination coefficient of the ionization was constant with values from 5×10^{-9} to 2×10^{-8} cm^3 sec^{-1}. Evidently, a high recombination coefficient produces a much sharper minimum in the electron density than that observed; while a low value produces a minimum which is much delayed from the time when the minimum electron density was observed. Many attempts were made to explain this discrepancy on the basis on an effective recombination coefficient which varied during the eclipse; for instance, Bates and McDowell (1957) suggested that two ions with quite different recombination coefficients might be present, resulting in an effective recombination coefficient that changed throughout the eclipse as the relative concentrations

of the two ions changed. Bowhill (1961) showed that one could obtain a reasonably
good fit to various E-region experimental results by assuming ions with recombination coefficients of about 6×10^{-9} and 6×10^{-8} cm^3 sec^{-1}. The ions were
tentatively identified as those of nitric oxide and molecular oxygen. However,
it is now known from laboratory measurements that the correct recombination
coefficients for these ions are about 5×10^{-7} and 2×10^{-7} cm^3 sec^{-1}, respectively, so obviously this explanation is not tenable.

    Some attempts have been made to reconcile the experimental results with a
single value of recombination coefficient by assuming that strong discrete
sources of radiation are present within the visible solar disc. However, such
analyses have nearly always concluded that very strong limb brightening is present, but with the western limb much brighter than the eastern. There seems to
be no physical reason why this would be expected; and, in fact, it can be shown
that this apparent result follows naturally from the assumption of a recombination coefficient large enough to prevent E-region ionization from disappearing
more rapidly than is observed during the period of totality. Evidently, it is
the neglect of residual radiation at totality which causes the difficulty.

    Taubenheim and Serafimov (1969) have established an interesting model for
the May 1966 solar eclipse in southern Europe, in which they assumed a coronal
contribution to the total ionizing intensity, proportional to the intensity of
the green coronal line. The recombination coefficient they estimate (greater
than 8×10^{-8} cm^3 sec^{-1}) is in much better agreement with laboratory measurements
than results typically obtained for solar eclipses.

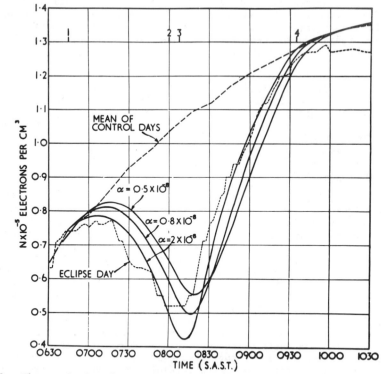

Figure 2.   Theoretical and experimental peak electron densities for the E layer
            (Szendrei and McElhinny, 1956).

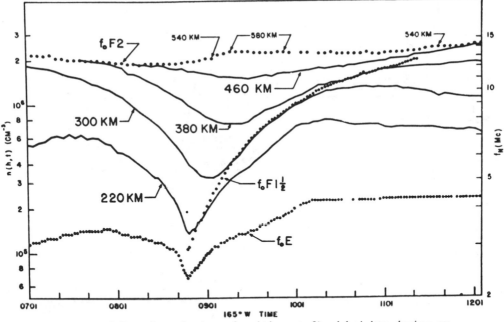

Figure 3. E- and F-region electron densities at fixed heights during an equatorial eclipse (Van Zandt et al., 1960).

Another argument in favor of the high values for recombination coefficient is the symmetry of the E-region electron density variation about the center of totality. A recombination coefficient of 5×10^{-7} cm^3 sec^{-1}, with a minimum electron density of 5×10^{-4} cm^{-3}, would yield a time delay of the minimum electron density relative to the center of totality of about 20 sec, scarcely detectable with normal experimental techniques. Lower recombination coefficients (say, 10^{-8} cm^3 sec^{-1}) would give about 17 min time delay, significantly greater than commonly observed (see, for instance, Figure 3).

If, then, we accept the position that the E-region ionization is essentially in photoequilibrium throughout even a total solar eclipse, it follows that the minimum value of the electron density should be due primarily to x-radiation from the sun. Since this radiation is substantially greater at solar maximum than solar minimum, it should be found that in a period of low sunspot activity the E-layer electron density decreases on the average by a larger ratio than in a time of high sunspot activity. It would be interesting to review past eclipses to see if this hypothesis is supported by E-layer critical frequency measurements.

It has been known for many years that the daily variation of the earth's magnetic field is affected by the passage of an eclipse (Bauer 1910). The effect is apparently due to a change in the conductivity of the E-region, produced by the decreased ionization in the eclipse path. Bomke et al. (1967) have used this effect to estimate the E-layer recombination coefficient during the November 1966 solar eclipse in Peru, by subtracting the diurnal variation of the magnetic field on a normal day from the eclipse day results. The difference field, shown in Figure 4, exhibited a time delay of nearly three minutes relative to the optimum location of the E-layer shadow, which he suggested might be due

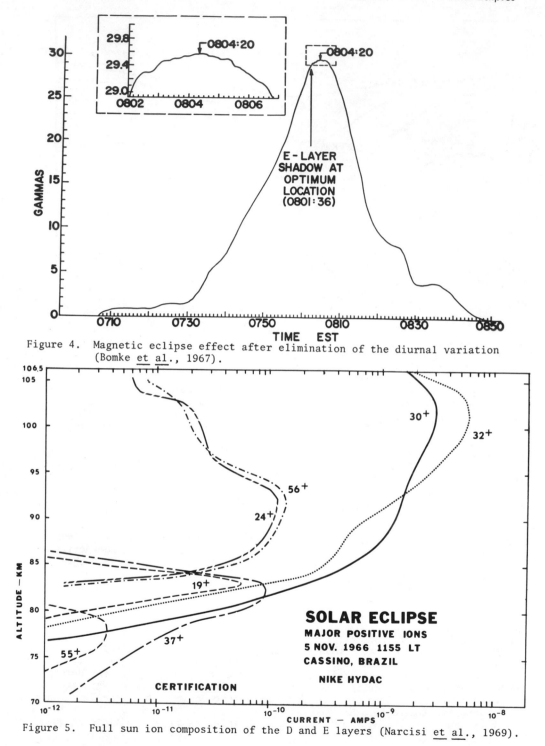

Figure 4.   Magnetic eclipse effect after elimination of the diurnal variation
(Bomke et al., 1967).

Figure 5.   Full sun ion composition of the D and E layers (Narcisi et al., 1969).

to a finite recombination coefficient for the E-region ionization. However,
Matsushita (1966) has pointed out that magnetic variations in solar eclipses
must be interpreted with caution, as the effect of the eclipse is to insert
a partial insulator in the path of the dynamo current system, and that the
dynamo current may be diverted around the obstacle; resulting in effects that
are hard to estimate quantatively.

Finally, there have recently been some measurements of ion composition
in the E region during a solar eclipse. Figures 5 and 6 (Narcisi et al. 1969)
show the relative abundances of various ions on a control day and during totality,
respectively, for the November 1966 total solar eclipse in southern Brazil.
While the totality measurements extended to only a little above 95 km, the
pattern of behavior is clearly that there is no substantial change in E-layer
ion composition during the eclipse. This lends further support to the idea
that photoequilibrium is maintained even at totality; resulting in a nearly
constant ratio of ion abundance, even though the recombination coefficients
may not be identical. Such change as is evident is in the direction of an
increase in the relative abundance of nitric oxide ions. Since, under full-
sun conditions, there is an approximately equal contribution from x-rays and
from the ultraviolet Lyman-β radiation, the fact that the eclipsed sun radiates
primarily x-rays, which ionize all atmospheric constituents (including atomic
oxygen and molecular nitrogen, which presumably contribute to the nitric oxide
ion concentration), would lead one to expect the relative abundance of molecular
oxygen ionization to decrease somewhat.

4. The F region during a solar eclipse

The key problem in the interpretation of F-region ionospheric results
during solar eclipses is that of distinguishing chemical from dynamic effects

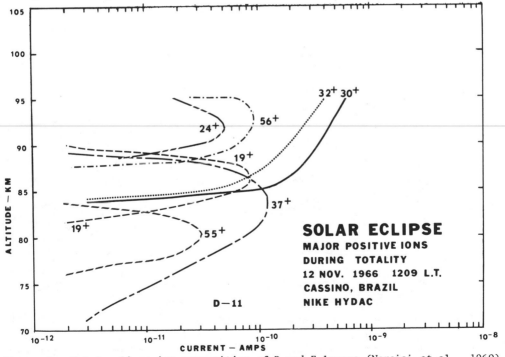

Figure 6. Total eclipse ion composition of D and E layers (Narcisi et al., 1969).

(Rishbeth 1968). If ionization movements can be neglected, the systematic variation of ionization production rates through a solar eclipse can give good measurements of the rate of recombination of F-region ionization, since the time constants involved (unlike those in the E region) are comparable with the eclipse duration.

Due to the effects of the earth's magnetic field, it happens that dynamic effects are much less important in the F region at low latitudes, where the magnetic field is nearly horizontal. The 1958 experiment of Van Zandt et al. (1960), reproduced here on Figure 3, illustrates the type of behavior that is typically encountered. The minimum electron density is delayed after totality by an amount which increases with altitude; and the depth of the minimum in electron density decreases correspondingly. Values of the electron loss coefficient deduced from these eclipse results are in good accord with recent laboratory measurements of ion-atom interchange rate coefficients, when combined with models of F-region molecular oxygen and nitrogen concentrations. Generally similar results were found by Skinner (1967) for the October 1959 eclipse in Nigeria.

A complication in F-region chemistry in solar eclipses is the observation that if the F1-layer is absent prior to the eclipse, it usually appears after the eclipse begins. As is well known, the altitude of the F1-layer maximum is close to the transition in the ion composition between the predominantly molecular ions of the E region and the predominantly atomic ions of the F region. De Jager and Gledhill (1963) have successfully explained this curious F1-layer behavior in terms of the change in character of the loss process at the altitude of the ion composition transition.

At medium latitudes, the behavior of the earth during an eclipse is quite different. Figure 7 shows a profile of the F2-layer electron density measured with the Thomson-scatter radar at Millstone Hill by Evans (1965). Relative

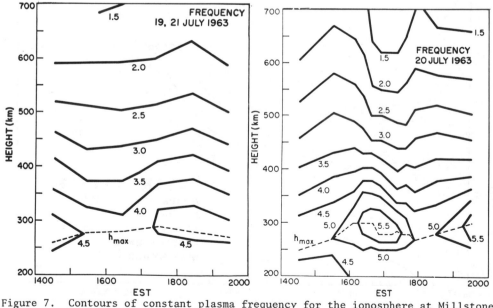

Figure 7. Contours of constant plasma frequency for the ionosphere at Millstone Hill (Evans, 1965). (a) control days   (b) eclipse day.

to the control days, the electron density on the eclipse day actually <u>increases</u>. Careful comparison of Figures 7(a) and 7(b) shows that the electron density has increased below 450 km altitude during the eclipse, but has decreased above that altitude. Evidently, therefore, the decrease in ionization production in this eclipse is relatively unimportant compared with transport effects; the ionization in the topside of the F layer has migrated down the field lines, resulting in an enhancement of the electron density at the F2 peak.

The reason for this rapid diffusion of topside ionization is shown on Figure 8, depicting the ratio of electron to ion temperature (also measured by the Thomson-scatter technique) on the control days and on the eclipse day. Whereas on a normal day the ratio is about 2, on the eclipse day it falls to as low as 1.2 (and it can be shown that the change involved is mostly in the electron rather than in the ion temperature). The distribution of ionization in the topside of the F layer is controlled by the plasma scale height, which in turn is determined by the sum of the electron and ion temperatures. Since this sum has decreased by approximately 30 percent during the eclipse, it follows that the plasma scale height must have decreased by a like factor, resulting in a sudden downward diffusion of ionization from the topside into the F2-layer peak; increasing it by the amount shown. On this basis, therefore, one would expect that a measurement of total electron content of the ionosphere would not show this increase; and, in fact, might be expected to decrease, if anything. Measurements in the same eclipse by Klobuchar and Whitney (1965) in fact showed a decrease in total electron content at the time of the eclipse.

This strong variation in the thermal structure of the ionosphere during an eclipse is due to the very rapid nature of the processes by which the ionization is heated (Geisler and Bowhill, 1965). Photoelectrons produced throughout the F region migrate along the lines of the earth's magnetic field, giving up energy by inelastic collisions to the neutral atmosphere, and by Coulomb interactions to the ambient ionization. If produced above 400 km, they may in fact escape

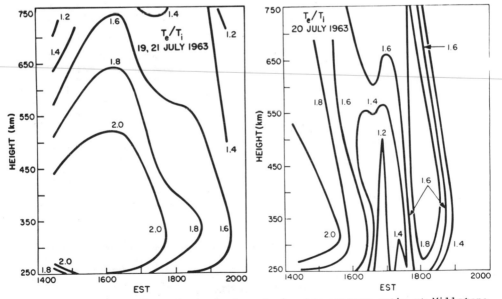

Figure 8.   Contours of constant electron-to-ion temperature ratio at Millstone Hill (Evans, 1965).   (a) control days   (b) eclipse day.

from the ionosphere and continue along the field lines to the geomagnetic
conjugate point. If the source of these photoelectrons is removed by a total
eclipse, it takes only a few seconds for the ionosphere to cool to a temperature
approaching the neutral gas temperature, the only remaining heat source being
conduction from the hot protonospheric ionization in the same field tube. The
ionization itself, of course, will react much more slowly to changes in thermal
structure, the rate being limited by the diffusion of ionization to the neutral
gas. However, it will be much more rapid at the highest altitudes. A proper
solution of the eclipse problem requires establishing the time-varying continuity
equation for the ionization, including transport effects due to changes in the
thermal structure. Work on this subject is only beginning (Cho and Yeh, 1970),
and has not yet been applied to the solar-eclipse case.

An interesting aspect of F-region eclipse effects is the possibility of
observing perturbations in the ionosphere at the point conjugate to that eclipsed
(Haubert and Laloë, 1963; Bousquet et al., 1967). One might think of an association between the contraction of the eclipse F2 layer, with the resultant drop
in plasma pressure at the top of the ionosphere, and a movement of ionization
from the conjugate hemisphere along the field lines, which would tend to reduce
the peak electron density at the conjugate point. The time taken for a com-
pressional wave to travel from one hemisphere to the other through the protonospheric plasma would imply a delay of several hours in the conjugate effect.
An alternative possibility arises from the eclipsing of the source of conjugate
photoelectrons. This would lead to a decrease in the heat input at the conjugate ionospheric point, coincident with the center of the eclipse; therefore,
presumably, a decrease in the electron temperature, and an increase in the
F2-layer maximum electron density. Present results do not seem to be conclusive
in this regard; perhaps a direct measurement of electron temperature at the
conjugate point would prove helpful.

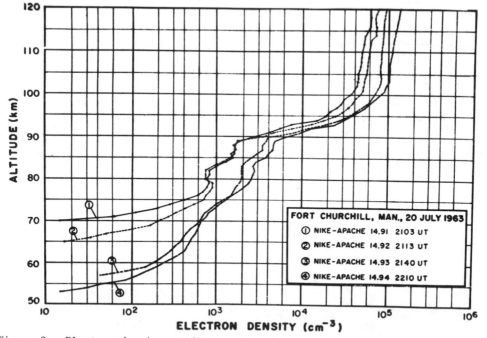

Figure 9.   Electron density profiles in the D and lower E regions assuming probe
current proportional to electron density (Smith et al., 1965).

5. The D region during a solar eclipse

The study of the D region during eclipse conditions, in common with the study of the normal D region, has suffered from the difficulty of measuring the very small electron densities (less than 10^4 cm^{-3}) which occur below 90 km. However, measurements of amplitude and phase of a VLF radio signal during an eclipse (Crary and Schneible 1965) in July 1963 showed a change in the phase height of reflection, with maximum excursion delayed a few minutes from the time of maximum obscuration at the center of the VLF path.

Smith et al. (1965) made direct rocket measurements of electron current collected by a Langmuir probe in the D region during a partial solar eclipse in July 1963 at Fort Churchill, Canada. The electron current curves from the four rockets are shown on Figure 9. Rocket 1 was fired just before maximum phase of the eclipse, when the visible disc was about 9 percent uncovered; rocket 2 ten minutes later, when the visible disc was about 14 percent uncovered; rocket 3, with the disc 59 percent uncovered; and rocket 4, with the disc 97 percent uncovered. Even though the proportionality constant between probe current and electron density may change with altitude below 90 km, it is obvious that the electron density at 70 km dropped dramatically during the maximum phase of the eclipse; by a greater factor, in fact, than even the area of the visible disc. We shall see later that this result is a key factor in interpreting the ion chemistry of this region.

During the November 1966 eclipse in southern Brazil, Mechtly et al. (1969) were able to measure electron densities with good absolute accuracy by a

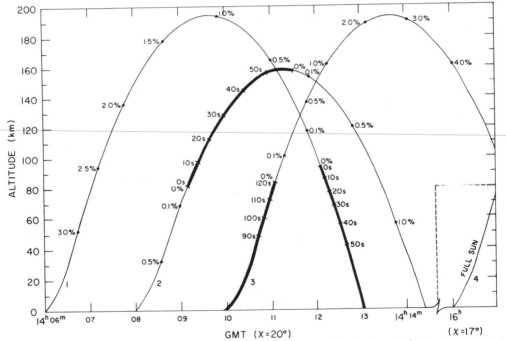

Figure 10. Percentages of visible solar disk, and elapsed times since commencement of totality at each position for four rockets launched in the November 1966 eclipse in Brazil (Mechtly et al., 1969).

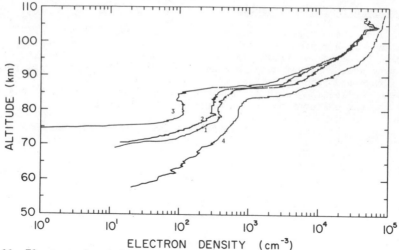

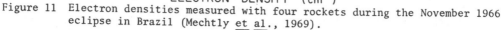

Figure 11 Electron densities measured with four rockets during the November 1966
eclipse in Brazil (Mechtly et al., 1969).

combination of the Langmuir-probe and radio-propagation techniques on a series
of four Nike-Apache rockets. The trajectories for the four rockets are explained
on Figure 10, and the electron density results are shown on Figure 11. Between
rockets 2 and 3, which were in the beginning and end of totality, respectively,
at 80 km altitude, the electron density dropped by a factor of three between
about 78 and 85 km, and by a much larger factor below 75 km. On the other hand,
the E-region ionization at altitudes about 90 km decreased by almost the same
factor at all altitudes, and did not change appreciably during totality.

Positive ion densities were measured by Bowling et al. (1967) using a
negatively biased probe on a series of sounding rockets during the solar eclipse
of May, 1966, in Greece. They found positive ion densities which decreased
from about 3000 cm^{-3} to about 300 cm^{-3} at 70 km altitude, between full sun and
totality.

Measurements of positive ion composition by Narcisi et al. (1969) are
shown on Figures 5 and 6. In contrast to the results of Bowling et al., no
very large decrease was found in the ion concentration at 80 km altitude;
though there was a greater relative abundance of doubly hydrated protons during
the measurements at totality.

In interpreting these results, Sechrist (1970) has suggested that electron-
ion recombination is the major loss process for electrons above 80 km altitude,
and has deduced a recombination coefficient of about 4×10^{-5} cm^3 sec^{-1} between
78 and 86 km from the results of Mechtly et al. (1969). This very large recombination coefficient is suggested to be associated with the recombination of
hydrated protons with electrons, measurements of which are not yet available
in the laboratory. It is interesting to note that both Bischoff and Taubenheim
(1967) and Crary and Schneible (1965) have deduced similarly large recombination
coefficients from VLF data.

Below 75 km, the extremely rapid disappearance of ionization can be explained
only by attachment, probably initially to molecular oxygen by a three-body
process, followed by charge exchange with ozone. Since atomic oxygen has the
ability, through associative detachment, to compete with the ozone for the avail-

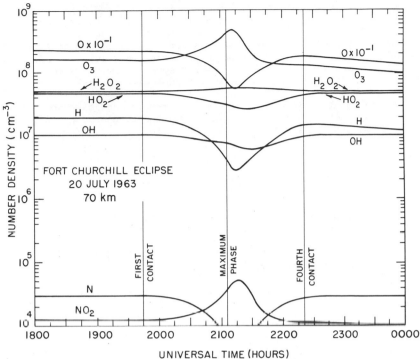

Figure 12. Minor constituent concentrations calculated for 70 km altitude, for
the 1963 eclipse at Fort Churchill, Canada (Keneshea et al., 1969)

able molecular oxygen negative ions, the rate at which negative ions of ozone
are formed probably depends critically on the atomic oxygen-to-ozone ratio.
Keneshea et al. (1969) have shown that this ratio can change dramatically during an eclipse; for example, Figure 12 shows their calculation of various minor
constituents at 70 km altitude for the circumstances of the partial eclipse at
Fort Churchill, Canada in July 1963. The ratio of these constituents changes
by more than an order of magnitude during this eclipse; and the effective loss
coefficient for electrons would therefore be expected to increase by a corresponding ratio. It is possible that the anomalous behavior of the electron density
below 75 km near totality could be explained on this basis; namely, a variation
in loss coefficient throughout the eclipse produced by the disappearance of
the dissociating influence of ultraviolet solar radiation from the visible disc.

6. Conclusion

    Interpretation of ionospheric measurements taken during solar eclipses,
in summary, will need to take into account the following considerations:

    1. The non-uniformity of solar radiation, both inside and outside the
        solar disc;

    2. The varying ion composition of the D and E region;

    3. Changes in electron and ion temperature in the F region, and the
        transport effects associated with them;

    4. Changes in minor constituent concentrations in the D region, and their effects on electron loss coefficients and ion chemistry.

In addition, it would be greatly advantageous to coordinate measurements using ground-based techniques such as sweep-frequency reflection sounding, satellite Faraday rotation, and Thomson-scatter sounding, with space techniques involving radio and direct sensing experiments on rocket vehicles.

It is certain that eclipse experiments will remain a highly useful tool for exploring transiant properties of ionospheric constituents, and determining rates for chemical and transport processes associated with them.

7. References

Bates, D. R. and M. R. C. McDowell (1957), Recombination in the ionosphere during an eclipse, J. Atmosph. Terr. Phys. 10, 96-102.

Bauer, L. A. (1910), The physical theory of the earth's magnetic and electric phenomena, Terr. Mag. Atmosph. Elect. 15, 107-128.

Beynon, W. J. G. and G. M. Brown (1956), Editors, Solar Eclipses and the Ionosphere, Pergamon Press, London.

Bischoff, K. and J. Taubenheim (1967), A study of ionospheric pulse absorption (A1) on 4 Mc/s during the solar eclipse of May 20, 1966, J. Atmosph. Terr. Phys. 29, 1063-1069.

Bomke, H. A., H. A. Blake, A. K. Harris, W. H. Hulse, D. J. Sheppard, A. A. Giesecke and A. Pantoja (1967), Recombination coefficient and coronal contribution to E-layer ionization from magnetic observations of a solar eclipse, J. Geophys. Res. 72, 5913-5918.

Bousquet, C., M. Lafeuille, G. Vasseur and P. Vila (1967), Some preliminary results of ionospheric observations during the eclipse of 20 May 1966, Ann. Geophys. 23, 345-356.

Bowhill, S. A. (1961), The effective recombination coefficient of an ionosphere containing a mixture of ions, J. Atmosph. Terr. Phys. 20, 19-30.

Bowling, T. S., K. Norman and A. P. Willmore (1967), D-region measurements during a solar eclipse, Planet. Space Sci. 15, 1035-1047.

Cho, H. R. and K. C. Yeh (1970), Neutral winds and the behavior of the ionosphere F2 region, Radio Sci. (in the press).

Crary, J. H. and D. E. Schneible (1965), Effect of the eclipse of 20 July 1963 on VLF signals propagated over short paths, J. Res. NBS D69, 947-957.

De Jager, C. and J. A. Gledhill (1963), The enhancement of the F1-cusp and the appearance of the F1 layer during solar eclipses, J. Atmosph. Terr. Phys. 25, 403-414.

Evans, J. V. (1965), An F-region eclipse, J. Geophys. Res. 70, 131-142.

Friedman, H. (1962), Solar observations obtained from vertical sounding, Reports on Progress in Phys. 25, 163-217.

Geisler, J. E. and S. A. Bowhill (1965), Ionospheric temperatures at sunspot minimum, J. Atmosph. Terr. Phys. 27, 457-474.

Haubert, A. and F. Laloë (1963), Ionosphère--une éclipse de soleil peut-elle modifier l'ionosphère au point magnétiquement conjugué?, J. Atmosph. Terr. Phys. 25, 105-107.

Keneshea, T. J. (1969), Theoretical variations of minor constituents during an eclipse, Aeronomy Report No. 32, University of Illinois, Urbana, Illinois, pp. 400-413.

Klobuchar, J. A. and H. E. Whitney (1965), Ionospheric electron content measurements during a solar eclipse, J. Geophys. Res. 70, 1254-1257.

Landini, M., D. Russo and G. L. Tagliaferri (1966), The solar eclipse of May 20, 1966 observed by the SOLRAD 8 satellite in the x-ray and ultra-violet bands, Nature 211, 393-394.

Matsushita, S. (1966), Effects of a solar eclipse on the equatorial geomagnetic field, Ann. Geophys. 22, 471-477.

Mechtly, E. A. and K. Seino (1969), Lower ionosphere electron densities measured during the solar eclipse of November 12, 1966, Radio Sci. 4, 371-375.

Narcisi, R. S., C. R. Philbrick, A. D. Bailey and Della Lucca (1969), Review of daytime, sunrise and sunset ion composition of the D region, Aeronomy Report No. 32, University of Illinois, Urbana, Illinois, pp. 355-363.

Rishbeth, H. (1968), Solar eclipses and ionospheric theory, Space Science Rev. 8, 543-554.

Sechrist, C. F., Jr. (1970), Interpretation of D-region electron densities, Radio Sci. (in the press).

Skinner, N. J. (1967), Eclipse effects in the equatorial F-region, J. Atmosph. Terr. Phys. 29, 287-295.

Smith, L. G., C. A. Accardo, L. H. Weeks and P. J. McKinnon (1965), Measurements in the ionosphere during the solar eclipse of 20 July 1963, J. Atmosph. Terr. Phys. 27, 803-829.

Taubenheim, J. and K. Serafimov (1969), Brightness distribution of soft X-rays on the Sun, inferred from ionospheric E-layer variations during an eclipse, J. Atmosph. Terr. Phys. 31, 307-312.

Van Zandt, T. E., R. B. Norton and G. H. Stonehocker (1960), Photochemical rates in the equatorial F2 region from the 1958 eclipse, J. Geophys. Res. 65, 2003-2009.

THEORETICAL ASPECTS OF SOLAR ECLIPSES

H. Rishbeth

Science Research Council, Radio and Space Research

Station Ditton Park, Slough, Bucks, U.K.

ABSTRACT

The variations of ionospheric electron density during an eclipse depend largely on the ionization loss coefficients and the photoionization rate. This paper reviews some factors which influence the analysis of eclipse observations of the E and F layers. The effects of electric fields, plasma diffusion and electron temperature changes on eclipse phenomena in the F2 layer are also discussed.

1.   INTRODUCTION

1.1   Object of the paper

A Symposium on solar eclipses and the ionosphere was held in London in 1955. The report of that meeting(Beynon and Brown,1956) serves as a full record of the subject as it existed at the time, and makes a convenient point of departure for a current review of eclipse effects in the E and F regions.

The present paper deals mostly with the difficult problem of obtaining reliable information about ionospheric processes from eclipse observations and not so much with the (perhaps easier)problem of obtaining information about the sun. It does not deal with the D region. Section 2 of the paper concerns what might be termed the "classical approach" to eclipse analysis, which uses the wellknown continuity equation for the electron density, and seeks to determine the rates of production and loss of ionization. Many of its results have been documented by Ratcliffe (1956a), in the report of the 1955 Symposium. It is disappointing to have to record that sub-

sequent progress in this field has not been extensive; some reasons
for this situation are discussed in section 3 of this paper. Some
eclipse effects related to transport processes, mostly in the F
region, are dealt with in section 4. Section 5 is a brief conclusion.

1.2   The normal daytime E and F layers

A summary of some important facts about the ionospheric layers
is appropriate here. The data refer to noon at middle and low latitudes; see for example Yonezawa (1966).

E layer. Peak electron density $NmE = (1-2)x10^5 cm^{-3}$,height
hmE $\approx$ 110 km. Produced by solar X radiation of wavelengths 10-170
Å approximately ; also UV radiation capable of ionizing O_2 but in-
capable of ionizing 0 and N_2, viz. 911-1027 Å. The rate of loss of
ionization is proportional to the square of the electron density,
namely αN^2, where α is the recombination coefficient.

F1 layer . Peak electron density $NmF1 = (2-4)x10^5 cm^{-3}$,height
hmF1 $\approx$ 170 km. Produced by solar UV radiation,wavelengths ~170-911
Å. The F1 layer lies at a transition between the domains of the
"square lay" loss formula αN and the "linear" loss formula βN;
this can account for the fact that the F1 layer is not always observable as a separate layer (Ratcliffe, 1956b).

F2 layer.   Peak electron density NmF2 usually in range $4x10^5-$
$-2x10^6 cm^{-3}$, height hmF2 in range 250-400 km. Produced by same radiation as F1, electron density being greater because the ration
of production rate to linear loss coefficient (q/β) increases upward. The peak occures at a level where transport processes,such
as diffusion, become more rapid than production and loss.Above
hmF2 the electron density is controlled by diffusion and decreases
upwards.

2.   CLASSICAL ECLIPSE ANALYSIS

2.1 The continuity equations

In what might be called the "classical" method of eclipse analysis, the time variation of electron density N(t) is interpreted
in terms of one of the continuity equations :

$$dN/dt = E(t)q(t) - \alpha N^2 + \{M\} \qquad (1)$$

$$dN/dt = E(t)q(t) - \beta N + \{M\} \qquad (2)$$

where {M} is the movement term, which is often supposed to remain
constant during the eclipse, and q(t) is the uneclipsed production
rate.

The eclipse function E(t) is often assumed equal to the geometrical fraction of the sun's disk not covered by the moon, in which case E(t) = 0 at totality, but this assumption is not necessarily true. As shown by Appleton (1953),these equations imply that changes of N lag behind changes of q with a time constant of order $1/(2\alpha N)$ for equation (1) and $1/\beta$ for equation (2).

Since effects in the E and F1 layers occur at the same time as the visible eclipse, these layers must be produced by solar photon radiation (Appleton and Chapman 1935). If the ionization were **due** to particles emitted from the sun, the interruption of the beam by the moon should occur some time before the visible eclipse For the F2 layer no such definite statement can be made, because the observed effects vary so much from one eclipse to another and because the layer is so prone to day-to-day and hour-to-hour fluctuations, though on other grounds it is not believed that particle effects are very important in the daytime F2 layer at mid-latitudes.

2.2 E layer results

By comparing the observed N(t) curves with the theoretical equation (1), it is possible to estimate q and α. During eclipses abrupt changes of dN/dt are often observed, and are attributed to the covering or uncovering of active regions on the sun (e.g.Minnis, 1958a).Often the more intense sources are found towards the limb, and it is thought perhaps 10% of the ionizing radiation originates in the solar corona and therefore is not out off at totality. Unfortunately, the determinations of α are extremely sensitive to the residual production at totality. Generally the assumption that q = 0 at totality leads to α 10^{-8} cm^3 sec^{-1}; but if the residual radiation is no more than 10-20%,values of 10^{-7}cm^{3}sec^{-1} or even more are obtained. These difficulties are further discussed in sections 3.1 and 3.2.

Sometimes only the variation of peak electron density NmE is analysed, and since this does not necessarily refer to a fixed height, extreme care must be used in interpreting the results.

2.3 F1 layer results

Very similar remarks apply to the analysis of F1 layer eclipse data. The values of α which are generally obtained are similar to those found for the E layer, perhaps slightly smaller,but as mentioned earlier the loss coefficient in the F1 layer is not expected to conform accurately to either of the equations (1) or (2). Sometimes the F1 layer appears during an eclipse,at a time or location where it would not normally be expected to be seen;or sometimes an additional stratification appears above the normal F1

layer. This is the socalled eclipse F1   layer,which is most often
seen in low magnetic latitudes. It is really no more than an infle-
xion on the electron density N(h) profile. Ratcliffe (1956a) has
explained how this phenomenon can arise in a layer where the loss
coefficient decreases upwards, as it does in the F region. A more
detailed study of F1   layer formation has been made by De Jager
and Gledhill (1963).

2.4 F2 layer results

Even under normal circumstances, the F2 layer is extremely
variable in its behaviour, so it is not surprising that a variety
of phenomena has been reported during eclipses. The reasons for
this variability may well be connected with movements of ionization which are thought to play an important part in the F2 layer.
Hence it would seem that useful information on production and loss
rates can only be obtained in circumstances when these movements
are comparatively unimportant. In low magnetic latitudes,largely
because of the hindrance to diffusion offered by the geomagnetic
field,eclipse behaviour in the F2 layer seems to be more regular.
Thus for the October 1958 eclipse,Van Zandt et al.(1960) obtained
production and loss rates, from N(t) curves at a series of fixed
heights, by plotting $(N^{-1}.dN/dt)$ versus E/N at frequent intervals
throughout the eclipse period. According to equation (2),this procedure should give a straight line graph from which q and β can
be determined, provided that {M} is unimportant and allowance is
made for the time variation of the normal production function q(t)
The success of this analysis seems to justify the assumption that,
for F2 layer production, the eclipse function E(t) is determined
by the geometrical obscuration of the solar disk.The deduced values
of q and β at 300 km were considerably greater than the values
that had previously been obtained by other means, but they are now
considered quite correct for the high level of solar activity pre-
vailing at the time of the eclipse (Rishbeth 1964).Skinner (1969)
derived similar values of q and β by a method that makes some
allowance for vertical drifts. Hence under favourable circumstances
eclipses may be capable of giving useful data on F2 production and
loss rates.

3.   DIFFICULTIES WITH THE CLASSICAL ANALYSIS

3.1 Residual production at totality

In section 2,2 it was remarked that values of recombination
coefficient deduced from eclipses may range from $^{\wedge}10^{-8}cm^3sec^{-1}$ to
$^{\wedge}10^{-7}cm^3sec^{-1}$,depending on whether the eclipse function E is assumed to be precisely zero at totality or to be of order 0.1. In
some cases, indefinitely large values of α can be made to fit
the data, providing correspondingly large values of q are adop-
ted. Hence the ionospheric observations do not suffice to deter-

mine q and α independently, and if eclipse data are to be of
any use in determining these parameters, it is essential to seek
further information about the uneclipsed radiation at totality.

In the normal E layer, there is a good correlation between
NmE and the flux of solar decimetric radiation (Minnis and Bazzard,
1959). According to radio astronomical measurements during a number of eclipses, summarized by Castelli and Aarons (1965),the residual solar flux at 10 cm wavelength is about 5-15%. Similar percentage residuals have been measured for X-rays of 44-60 $\overset{\circ}{A}$ - a
possible source of E layer ionization - by rocket experiments during eclipses (Smith et al.1965).These results are consistent
with E layer recombination coefficients of order $10^{-7} cm^3 sec^{-1}$,
such values being also favoured by laboratory plasma experiments,
which do not support values as low as $10^{-8} cm^3 sec^{-1}$ (Biondi 1964).
But it does not seem prudent,without further evidence,to assume
a detailed correlation between production rates and solar 10 cm
flux throughout an eclipse,even though such a correlation exists
on a day-to-day basis.

3.2   Localized sources of radiation

An important use of ionospheric eclipse observations is the
study of the distribution of the sources of ionizing radiation.
If data from a number of stations are available, then the sources
of radiation can be located quite accurately. For instance, the
radiation which produces the E layer is found to be largely concentrated in active areas,generally situated at the limb of the sun
(e.g.Minnis, 1958b).Even for this purpose,however, the ionosphere
is not a perfect detector. This is because,according to the continuity equations (1) and (2), the covering or uncovering of an active region changes only dN/dt; it does not immediately after N.
The information obtainable from ionospheric eclipse data does not
really compare in quality of detail with what can be learned from
X-ray photographs obtained from rockets, which do not require an
eclipse (Underwood and Muney,1967).

3.3   Complex photochemistry of the ionosphere

The "classical" eclipse analysis based on equ.(1) assumes that
a single recombination coefficient α exists in the ionosphere.
Bates and McDowell (1957) pointed out that certain E region phenomena, including some eclipse effects, might be explained by the
existence of two species of positive ions possessing different
recombination coefficients. In fact, rocket mass spectrometer experiments (Johnson et al.1958) have shown that two species of ion
NO^+ and O_2^+, predominate in the E region.But the recombination
coefficients for these two ions are probably not very different
nor does Minnis (1958b) consider the two-ion theory adequate to

explain the data.

It is clear that information about the ion composition is essential to any detailed study of the E and F1 layers. At these
heights, such information is normally obtained by the use of rocket
borne mass spectometers. Some data on the (molecular ion/atomic ion)
concentration ratio can be derived from incoherent scatter measurements though not, as yet, during eclipses.

4.   TRANSPORT EFFECTS DURING ECLIPSES

4.1. Ionospheric electric currents and associated geomagnetic varia-
tions

This subject was discussed at the 1955 Eclipse Symposium by
Chapman (1956).Changes of electron density during an eclipse must
alter the electrical conductivity of the ionosphere.This may affect the distribution of electric field and will certainly alter
the electric currents, which flow mainly in the E layer. Changes
of current will produce slight changes in the geomagnetic fiels,
such as are observed at ground level.A study of these changes during the November 1966 eclipse enabled Bomke et al.(1967) to derive an estimate of the E layer,recombination coefficient,
$\alpha = 5.5 \times 10^{-8} cm^3 sec^{-1}$.

4.2   Electromagnetic movements

According to the theory of Martyn (1955),electric fields generated in the E region by dynamo action cause electromagnetic
movements in the F2 layer. These movements are believed to be responsible for the production of the well-known F2 layer equatorial
anomaly. This explanation requires an upward drift at the equator
by day. The eclipse of November 1966 enabled this drift to be detected by the incoherent scatter technique at Jicamarca. A series of
N(h) profiles, obtained by V.L.Peterson and others,showed that the
eclipse punched a "hole"in the electron distribution at around the
production peak in the F1 layer; this "hole" subsequently drifted
upwards at about 50 m sec^{-1}. This is faster than (but in the same
direction as) the drift required to produce the equatorial anomaly
on normal days, and therefore it is necessary to consider how the
electric field might have been affected by the eclipse. It is
equally necessary to investigate how the existance of this drift
might influence the determinations of q and β described in
Section 2.4. Skinner (1969) has done this (see Sec.2.4);but when
applied to N(h) data for the October 1959 eclipse in West Africa,
his method gave downward drifts of 5-50 m sec^{-1}. These disagree
seriously with the upward drifts measured at Jicamarca.

4.3. Plasma diffusion during eclipses

The F2 layer at mid-latitudes is thought to be largely control-

led by plasma diffusion,which determines the height at which the
F2 peak is found. Theoretically, diffusion tends to smooth out differences in the behaviour of the layer at different heights.Below
the F2 peak, where diffusion is comparatively slow, the eclipse
effect at any height should be governed by the local value of the
loss coefficient β ,according to equation (2).The coefficient β
decreases upwards, and loss is negligible at heights well above the
peak;but plasma can readily diffuse downwards to replace the loss
by recombination at lower levels. As a result the variations of
electron density in the topside of the F2 layer should be rather
similar to those at the F2 peak (Rishbeth 1963).Eclipse effects
observed with the topside sounder (King et al.1967) seem generally
consistent with this behaviour,although part of the observed reduction is probably due to the thermal contraction of the plasma,dis-
cussed in section 4.4.

        Since diffusion only takes place along magnetic field lines,
these considerations do not apply in low magnetic latitudes and the
behaviour of the layer is dominated by production and loss,even at
great heights.

4,4   Reduction of electron temperature

        Another process which is controlled by the direction of the
geomagnetic field is the conduction of heat by ionospheric electrons.
It is well established that in the daytime F layer $T_e > T_n$ although $T_i \simeq T_n$, where T_e, T_i, T_n denote respectively the electron,
ion and neutral gas temperature. The excess electron temperature
results from the large amount of energy acquired by newly libera-
ted photoelectrons, which have poor thermal contact with the heavier particles. If photoionization is interrupted by an eclipse, T_e
may fall rapidly, as the excess heat is drained away by conduction
to lower heights. Such a reduction has been observed by Evans(1965a).
Since in some respects the F region plasma behaves as a gas possessing a temperature $\frac{1}{2}(T_e + T_i)$, a reduction of T_e tends to cause
thermal contraction of the plasma. If sufficiently rapid, this contraction may lead to an increase in F2 peak electron density,such as
sometimes occurs during eclipses. It should also cause a downward
movement of the topside of the F region.Leading to a marked local
reduction of electron density, which has also been observed (King
et al.1967).Evans (1965b) has studied data from many past eclipses
and found that increases of NmF2 are observed whenever (i) the eclipse is virtually total (> 90%)in the F1 layer, at which height
the input of energy to the electrons is greatest; and (ii) the magnetic dip angle is fairly large (> 60%),to facilitate the loss of
heat by conduction to lower heights. These are favourable conditions
for producing a large reduction of electron temperature.

5.   CONCLUSION

     Ionospheric eclipse observations can make a worthwhile contri-
bution to study of the sun, by giving information about the location
of sources of ionizing radiation. It is however not easy to derive
quantitative information on processes in the ionosphere.

     In the E region and the lower F region, production and loss
processes are thought to dominate the variations of electron density. The idealized continuity equations, given in section 2.1, should
hold approximately; but it is difficult in practice to obtain useful information about the photochemical rates from eclipse observations. The difficulty can be attributed to the questions of uneclipsed radiation at totality (section 3.1) and the complex photochemistry of the ionosphere (section 3.3). These questions can only be
resolved by the use of **additional techniques,** notably rocket-borne
experiments;for instance, mass spectrometers can give information
on ion composition. But it is not yet known whether such experiments during eclipses can give any distinctive information about the
ionosphere, which is not obtainable by the same experiments conducted under normal circumstances. The basic problem remains, that one
is using a variable source (the eclipsed sun) to study a complex
system (the ionosphere), whose response time is comparable to the
time scale of the eclipse. On the other hand, there is a possible
advantage in that, during midday eclipses, the production rate varies rapidly while the zenith angle changes only slowly; i.e.$q(t)$
in equations (1) and (2) is slowly-varying. This minimizes the
problems of calculating optical depth such as occur in the interpretation of ionospheric changes at sunrise and sunset. Incidentally,
it might be hoped that particularly interesting information could
be obtained from observing an eclipse near sunrise; but although
the February 1961 eclipse offered such an opportunity in Europe,
this hope does not seem to have been realized.

     Contrary to what might be expected, the most promising recent
determinations of production and loss rates during eclipses have
been made in the F2 layer, but only in equatorial latitudes where
vertical diffusion is slow. However, the Jicamarca observations
mentioned in section 4.2 show that electromagnetic drifts are present during an eclipse, and it remains to be demonstrated that
these drifts do not unduly influence the determination of q and
β . More observations of this type, and further theoretical calculations, would be valuable. The thermal effects during an F-region
eclipse, described in section 4.4, should provide scope for further study.

ACKNOWLEDGMENTS

This paper is published by permission of the Director of the Radio and Space Research Station in the U.K.Science Research Council. It is essentially the text of a review lecture presented at the NATO Advanced Study Institute in Athens,May 1966,with some later revisions. An expanded review has been published by Rishbeth (1968).

REFERENCES

Appleton E.V.and Chapman S. *(1935) Proc.I.R.E.23,658*

Appleton E.V. *(1953) J.Atmos.Terr.Phys. 3,282*

Bates D.R.and McDowell M.R.C. *(1957) J.Atmos.Terr.Phys.10, 96*

Beynon W.J.G. and Brown G.M. *(1956) Solar Eclipses and the Ionosphere, Pergamon Press,*

Biondi M.A. *(1964) Annales de Geophysique 20,34*

Bomke H.A.,Blake H.A., Harris A.K.,Hulse W.H., Sheppard D.J.,Giesecke A.A. and Pantoja A. *(1967) J.Geophys.Res.67,5913*

Castelli J.P. and Aarons J. *(1965) Chapter 3 in Solar System Radio Astronomy,Plenum Press,N.Y.*

Chapman S. *(1956) Solar Eclipses and the Ionosphere,Pergamon Press,Lon.p.221*

De Jager C. and Gledhill J.A. *(1963) J.Atmos.Terr.Phys.25,403*

Evans J.V. *(1965a) J.Geophys.Res.70,131*

Evans J.V. *(1965b) J.Geophys.Res.70,733*

Johnson C.Y.,Meadows E.B. and Holmes J.C. *(1958) J.Geophys.Res.63,443*

King J.W.,Legg A.J. and Reed K.C. *(1967) J.Atmos.Terr.Phys.29,1365*

Martyn D.F. *(1955) Physics of the Ionosphere, Physical Society,London,p.254*

Minnis C.M. *(1958a) J.Atmos.Terr.Phys.12,266*

Minnis C.M. *(1958b) J.Atmos.Terr.Phys.12,272*

Minnis C.M. and Bazzard G.H. *(1959) J.Atmos.Terr.Phys.14,213*

Ratcliffe J.A. *(1956a) Solar Eclipses and the Ionosphere,Pergamon Press,Lon.p.1*

Ratcliffe J.A. *(1956b) J.Atmos.Terr.Phys.8,260*

Rishbeth H. *(1963) Proc.Phys.Soc.81, 65*

Rishbeth H. *(1964) J.Atmos.Terr.Phys.26,657*

Rishbeth H. *(1968) Space Sci.Rev.8,543*

Skinner N.J. *(1969) J.Atmos.Terr.Phys.31,1333*

Smith L.G.,Accardo C.A.,   *(1965)*   *J.Atmos.Terr.Phys.*$\underline{27}$*,803*
    Weeks L.H.and McKinnon P.J.
Underwood J.H.and   *(1967)*   *Solar Physics* $\underline{1}$*, 129*
    Muney W.S.
Van Zandt T.E.,Norton R.B.   *(1960)*   *J.Geophys.Res.*$\underline{65}$*,2003*
    and Stonehocker G.H.
Yonezawa T.   *(1966)*   *Space Sci.Rev.* $\underline{5}$*,3*

THE EARTH'S EXOSPHERIC PLASMA WITH SOME COMMENTS ON

ECLIPSES AS A MEANS OF STUDYING THE IONOSPHERE

J. O. Thomas

Physics Department, Imperial College

University of London

Introduction

In this paper I have been asked to describe briefly the main factors governing the physics of the earth's exospheric plasma distribution so that the effects of solar eclipses can be related to the relevant geophysical theoretical background. F region chemistry is discussed elsewhere in this volume and the ideas described in this paper apply to the topside of the ionosphere, above an altitude of about 600 km. There, to a first order, theory and experiment agree that diffusive equilibrium of the plasma in the presence of the earth's magnetic and gravitational fields is the dominant physical process deciding the plasma distribution. As a preface I should, however, mention some observations made during the eclipse of 15th February 1961 with which I have been concerned and which are described in more detail in a companion paper (Thomas and Rycroft, this volume). In addition, some general precursive comments are made on eclipses as a tool for exospheric and ionospheric plasma studies.

The solar eclipse of 15th February 1961 was interesting in two respects. First, because the eclipse occurred close to the sunrise period - a very rare event. At several of the observatories the sun actually rose eclipsed - providing an extension to the normal night. Secondly, the eclipse effects were monitored at several fairly closely grouped stations, including Athens, Ebro, and Paris. Thus one can look at the geographical relationships fairly accurately and observe the size of the zone in the ionosphere over which the conductivity is reduced. There is a need for the kind of calculation in which the volume over which the conductivity

is reduced is estimated theoretically and compared with such observations. The way in which the ionization returns to normal within
this zone can then be examined. In particular the effects of movements accociated with eclipses might be assessed using this aproach.
With the very high density of stations in Europe, it might be possible to look for eclipse effects outside the eclipsed zone.

     In general it is worth bearing in mind that eclipses provide
a good way of determining the ratio of the loss rates at different
altitudes in the F region. The equalising effects of diffusion in
the topside ionosphere makes eclipse applications difficult. However, near the equator where the field lines are nearly horizontal,
vertical diffusion of ionization is inhibited, therefore the effects
of eclipses at high and low latitude observatories is likely to be
very different in character and indeed, this is found to be the
case. Because of the different role of diffusion in the two cases,
the effects of production and loss should be more important to
greater altitudes near the equator.

     Changes in electron temperature are also important. Evans
has pointed out that high latitude eclipses may show up these
changes more than low-latitude records since it is relatively
easier to remove heat at high latitudes so that the T_e changes will
be greater. There will then be a corresponding change in the scale
height – the decrease in the ratio T_e/T_i during an eclipse resulting in a decrease in scale height. Thus an enhancement (due to
downward drift of the plasma along the field lines) of the ionization at the F region peak, may result instead of the expected
decrease during an eclipse.

     Finally attention should be drawn to the need to record magne-
tograms and to examine micropulsation effects during an eclipse.
Kato in Japan has shown that the vector diagram showing the NS and
EW components change markedly during an eclipse presumably because
the change in E region conductivity affects the propagation of
hydromagnetic waves.

Physics of the exospheric plasma

     There are at least four zones which can be delineated in terms
of the basic physics which applies there.

1.   The first zone may be called the outer plasmasphere. This
zone lies between the whistler knee or plasmapause and a lower
limit in the ionosphere called ' the base of the neutral exosphere.'
The latter lies at about 600 km altitude, the exact location being a
function of solar activity. It is defined as the level where the
mean free path of the neutral particle is equal to the scale height.

At high latitudes the plasmapause is often detectable as a sharp
decrease or trough in the F region electron density. Inside the
plasmasphere chemical effects are not the dominating process. Thus
one can think of this region as starting at about 600 km and extending out to 3 or 4 earth radii: within it, diffusive equilibrium
applies to a first order.

2.   The second zone may be called the lower plasmasphere. It can
be defined as that part of the earth's plasma envelope lying below
the base of the neutral exosphere. It thus contains the main ionospheric layers E, F1 and bottomside F2. In the lower plasmasphere
production and loss effects, ion chemistry and various transport
phenomena dominate the physics.

3.   The third region lies outside the plasmasphere proper and it
is very likely that there, diffusive equilibrium breaks down. The
property which characterises this part of space is that here particles are lost from the atmosphere. Particle trajectories are important. If the particle is in an elliptic orbit, it will return to
the earth's atmosphere and will continue to be a part of the earth's
atmosphere. On the other hand, if the trajectory of the particle,
now in a collisionless region, is hyperbolic, then the particle (be
it a neutral particle or perhaps an ion) will probably be lost from
earth's atmosphere. In this region one cannot apply the basic ideas
of diffusive equilibrium, one has to look at these ballistic trajectories and apply the kind of analysis described by Eviator, Lenchek
and Singer (1964).

4.   The fourth zone is the polar exosphere which is characterised
essentially by dynamics. The main difference between the current
theoretical ideas about the diffusive equilibrium region above the
exospheric base and the polar exosphere lies in the fact that the
diffusive equilibrium solution is essentially one where the velocity
is zero. There may be a time required to set up diffusive equilibrium, but once that has been done essentially the solutions
correspond to the case where V = O. In the polar region, one has,
for a number of reasons a situation where it is very likely that at
least some of the positive ions in the top of the ionosphere are
moving with relatively large velocities. In the case of hydrogen
ions and helium ions, these are probably supersonic velocities.
Under these circumstances, the diffusive equilibrium equations have
to be modified to include the term which represents the momentum of
relatively fast moving heavy ions. In the diffusive equilibrium
picture one puts that term equal to zero.

The polar region is further complicated by the fact that
there are influxes of particles along field lines. Furthermore,
recent work by King and his colleagues at Slough shows that, in
summer, most of the rather peculiar characteristics of the F region

behaviour can be explained on the basis of neutral winds (derived from density observations made by satellites) modifying the basic plasma distribution. Clearly in the polar region, beyond the plasmasphere termination one is essentially in an atmosphere where the dominating terms are those of plasma dynamics.

Basic equations for the plasma distribution in the exosphere under diffusive equilibrium

The basic equations are those for the distribution of plasma in diffusive equilibrium under gravity, in the presence of the earth's field. A complete exposition of the relevant theory (which applies to the approximate height range, 1,000 -32,000 km ($\sim$5 earth radii)) is given by Angerami and Thomas (1964) and in the papers referenced therin. The main assumptions on which the theory is based are the following:
(1) The upper atmosphere above 1,000 km consists of neutral particles together with a neutral mixture of singly charged positive ions and electrons only, and these are in diffusive equilibrium. The positive ions are O^+, He^+ and H^+. (2) The partial pressure for each species is balanced by the earth's gravitational and centrifugal forces and the electric field arising from charge separation. (3) The charged particles are constrained to move only along the lines of force so that the distributions along different lines of force are quite independent of each other. (4) No electrons are produced by the action of the Sun's ionizing radiations above 1,000 km. (5) The rate at which electrons recombine is so small in comparison with other effects that the loss of electrons from this cause can be neglected in the calculations. (6) The earth's magnetic field is a dipole field, and the axis of rotation of the earth coincides with the magnetic dipole axis. The following assumptions are not inherent in the theory but may be made in order to simplify the calculations. (7) The electrons and ions have the same temperatures ($T_e = T_i$). (8) The ion temperatures (and therefore the electron temperatures are constant along a magnetic field line above 1,000 km.

In the nomenclature used by Angerami and Thomas (1964), (and used throughout this article), assumptions 7 and 8 above correspond to the conditions $C = T_{e_0}/T_{i_0} = 1$ and $T(s)/T_0 = 1$, respectively. Here the subscript zero refers to quantities measured at a reference-level at 1,000 km above the earth and s is a distance measured along a line of force from its foot at the reference-level.

The electron density n_e as a function of distance, s, along a line of force is given by:

$$\frac{n_e(s)}{n_{e_o}} = \left\{ \frac{1}{\mu} \Sigma \left[\mu_i \exp \frac{-z}{H_i} \right] \right\}^{\frac{1}{2}} \tag{1}$$

where n_{e_o} is the electron density on the line of force at the reference-level at the base of the exosphere, H_i is the scale height of oxygen, helium, and hydrogen atoms for $i = 1,2,3$, respectively, and μ_i is the number density of each positive ion normalized to the O^+ number density, that is, $\mu_i = n_i/n_1$. Note that:

$$n_e = (1 + \mu_2 + \mu_3)n_1 = \mu\, n_1$$

so that the quantity μ_i/μ which appears in equation (1) is merely the fractional abundance of the ith ion in the mixture.

In equation (1) the quantity z for a point at a distance s' along a field line is defined by:

$$z = \int_{s=0}^{s=s'} \frac{f}{g_o} \frac{T_o ds}{T(s)} = \int_{s=0}^{s=s'} \frac{f}{g_o} ds \tag{2}$$

in which f is the total force acting on an ion and g_o is the acceleration of gravity at the base-level. The quantity z at the apex of the field line in the equatorial plane can be calculated from the relation:

$$z = r_o \left[1 - \cos^2\theta_o + \frac{\Omega^2 r_o}{2g_o} \left(\cos^2\theta_o - \frac{1}{\cos^4\theta_o} \right) \right] \tag{3}$$

(equation (B.5) ibid.). Here r_o is the distance from the centre of the earth to the reference-level, θ_o is the geomagnetic latitude of the particular field line at the reference-level, Ω is the angular velocity of the earth's rotation about its geographic axis, and g_o is the accerleration due to gravity at the reference-level.

Underlying this theory are the basic momentum conservation equations for the electrons and ions in diffusive equilibrium under gravity in the presence of the earth' field. The theory involves an electric field directed upwards along the lines of force such that it prevents electrons from moving upwards away from the positive ions. When $T_e = T_i$ the electric field corresponds to a force $m_i g/2$ where m_i is the ionic mass. For O^+ this is $8g$ so that a proton would experience a net upwards force of $7g$. The field lines are closed. The upwards E field in each hemisphere decreases along the line of force, and there is no net field in the equatorial plane. (Note that when the field lines are open as in the polar wind case, the upwards acceleration of protons is large and super-sonic velocities are achieved). The reader is referred to the paper by Angerami and Thomas (1964) for a detailed discus-sion of the ion and electron distributions along field lines pre-dicted by this theory. A very large amount of experimental obser-vation both from whistlers and in-situ measurements during recent years confirm that over the height range referred to earlier, the theory adequately predicts the main variations. It is important to note that in these circumstances, diffusion recovery times are relatively fast compared with times like $1/\beta$ or $1/2\alpha N$.

Scale height, mean ionic mass and temperature variations

In terms of the theory and assumptions outlined above it is the scale height of the field-aligned plasma, H_S, rather than the scale height of the corresponding vertical electron distribution, H_V, which is physically significant where

$$H_V = -N/\frac{\partial N}{\partial r} \qquad \lambda = \text{constant}, \tag{4}$$

$$H_S = -N/\frac{dN}{dr} \qquad L = \text{constant}. \tag{5}$$

At high latitudes where the dip angle is large, H_V is a good approximation to H_S. At low latitudes the difference between the two quantities is appreciable. The pertinent relation may be writ-ten (Chandra and Goldberg 1964, Goldberg 1965, Thomas and Rycroft 1970)

$$\frac{\overline{m}_i g}{k(T_e + T_i)} = \frac{-1}{N}\frac{\partial N}{\partial r} + \frac{\cot \lambda}{2Nr} - \frac{1}{T}\frac{\partial T}{\partial r} + \frac{\cot \lambda}{2Tr}\frac{\partial T}{\partial \lambda} \tag{6}$$

or

$$\frac{1}{H_S} = \frac{1}{H_V} - \frac{A}{H_\lambda} + \frac{1}{G_V} - \frac{A}{G_\lambda} \tag{7}$$

where

$$H_S = \frac{k(T_e + T_i)}{\overline{m}_i g} \tag{8}$$

and

$$\frac{1}{H_V} = -\frac{1}{N}\frac{\partial N}{\partial r} ; \qquad A = \tfrac{1}{2}\cot \lambda ; \qquad \frac{1}{H_\lambda} = -\frac{1}{Nr}\frac{\partial N}{\partial \lambda} \text{ etc}$$

The G_V and G_λ terms representing the vertical component of variations of the effective plasma temperature, T, and the latitudinal gradients of T are such that above 800 km they may be set equal to zero to a first approximation. H_V and H_λ are the vertical and latitudinal components of the scale height of the plasma distribution. The nomenclature is summarized in Table 1.

Table 1
Nomenclature

g	acceleration due to gravity
h	altitude defined as $(r - R_E)$
k	Boltzmann's constant
m_e	electron mass
$\overline{m}_i$	mean ionic mass
r	radial distance from the earth's center
s	distance along a field line
H, H_V	vertical plasma scale height, defined as $-N/(\partial N/\partial h)$
H_S	scale height of field-aligned plasma defined as $-N\|(dN/dr)\|_{L=constant}$
I	magnetic dip angle
L	McIlwain's parameter; defining a line of force of the geomagnetic field
N, n_e	electron density
R_E	radius of the earth
T_e	electron temperature
T_i	ion temperature
T	effective temperature of the plasma, defined as $\frac{1}{2}(T_e + T_i)$
λ	dip latitude, defined as arctan $(\tan \frac{1}{2}I)$

Variations of H_S with latitude may be interpreted on the diffusive equilibrium model as being latitudinal variations of the ratio of the sum of electron and ion temperatures to the mean ionic mass (equation 8).

Topside and bottomside soundings (h'(f) curves) are often used to derive vertical N(h) profiles from which a scale height is deduced. It is important to note that this scale height is H_V (not H_S). High altitude H_V plots against latitude are characterised by a very large equatorial bite-out, Fig 1 (even when the corresponding N curve shows no bite-out, Fig 2). This feature (Fig 1) is the main indication of geomagnetic control in the upper equatorial F region. To derive $\overline{m}_i$ (given T_e and T_i) or T (given $\overline{m}_i$) within $\pm 25°$ latitude north and south of the equator it is extremely important first to convert the observed H_V into H_S, the field aligned plasma scale height using (7) before applying (8) (Thomas et al. 1966). At higher latitudes $H_S \simeq H_V$.

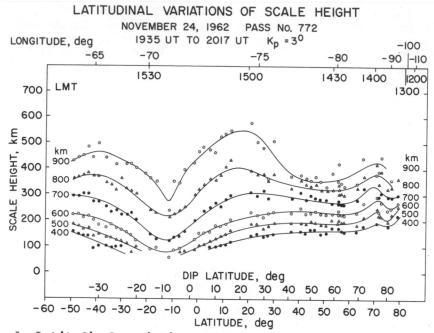

LATITUDINAL VARIATIONS OF SCALE HEIGHT
NOVEMBER 24, 1962   PASS No. 772
1935 UT TO 2017 UT $K_p = 3^0$

Fig. 1   Latitudinal variations of plasma scale height of vertical
electron distribution (H_v) on 24 November 1962. Latitudes
south of the equator and longitudes west of the Greenwich
Meridian are preceded by a minus sign, (from Thomas et al.
(1966)).

Bearing the above in mind, we now drop the suffixes and write
H for the scale height of a spherically symmetric plasma in diffusive equilibrium. Then differentiating (8)

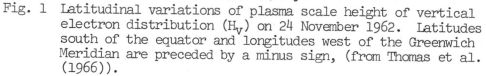

$$\frac{1}{H}\frac{dH}{dh} = \frac{1}{(T_e + T_i)}\frac{d(T_e + T_i)}{dh} - \frac{d\bar{m_i}}{dh} - \frac{1}{g}\frac{dg}{dh}. \qquad (9)$$

Since $d(T_e + T_i)/dh < + 200^0K/100$ km above 600 km at most local
times then, with $t \sim 2000^0K$, the first term in (9) is $\sim + 5\%/100$ km.
Using the inverse square variation of the earth's gravitational
field with distance, the last term in (9) is found to be
$\sim - 2\%/100$ km in the topside ionosphere. Hence variations in
$(1/H)(dH/dh)$ which are greater than $\sim + 7\%/100$ km are explained by
decreasing values of $\bar{m_i}$ with increasing altitude. dH/H is usually
considerably greater than 7% between two levels separated by
100 km. Thomas and Rycroft (1970) have used a temperature model
based on satellite observations and on incoherent scatter data to
deduce $\bar{m_i}$ using (7) and (8). The model is reproduced in Figs. 3
and 4 and is characteristic of quiet sun conditions. The results
may be summarized by stating that above an altitude of 600 km,

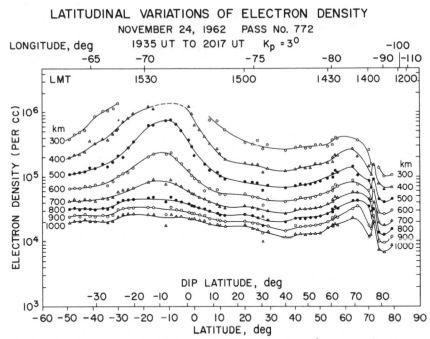

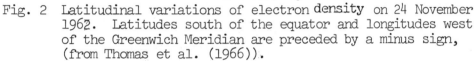

Fig. 2   Latitudinal variations of electron density on 24 November
         1962.   Latitudes south of the equator and longitudes west
         of the Greenwich Meridian are preceded by a minus sign,
         (from Thomas et al. (1966)).

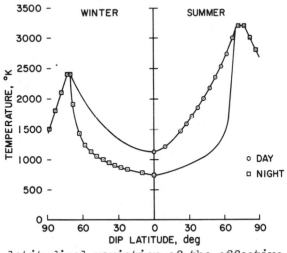

Fig. 3   Model latitudinal variation of the effective temperature
         of the ionospheric plasma above 800 km by day and by night,
         constructed from several sources, (from Thomas et al.
         (1966)).

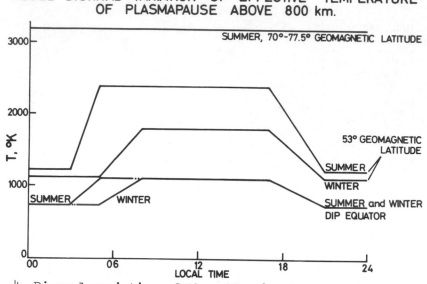

MODEL DIURNAL VARIATION OF EFFECTIVE TEMPERATURE
OF PLASMAPAUSE ABOVE 800 km.

Fig. 4   Diurnal variation of the effective temperature, T, of the
         exospheric plasma between 800 and 1000 km (from Thomas and
         Rycroft (1970)).

vertical temperature gradients are, except near local sunrise and
sunset, less than $1^{\circ}K\ km^{-1}$. Daytime ion temperatures, ~$1300^{\circ}K$ at
mid-latitudes, are <1.5 times greater than the night-time values.
The ratio of electron to ion temperature is generally ~1.2 by night,
and greater (~1.6) durnig the day, being ~3 at sunrise in mid-
latitudes. Both electron and ion temperatures increase with
increasing geomagnetic latitude towards the auroral zones. Fig. 3
refers to day and night conditions and gives the effective temperature of the plasma as a function of latitude. Fig. 4 presents
a realistic model of the diurnal variation of effective temperature
T, derived from Jicamarca incoherent-scatter observations at the
equator and from Evans' (1965) M.I.T. results at higher latitudes.

        It is found that the mean ionic mass around midnight at a
low solar epoch is typically 2 or 3 AMU near 50° latitude at
1000 km indicating the predominance of He^{+} and H^{+} ions in the
plasma. At sunrise the ionosphere is heated and expands so that
$\bar{m}_i$ increases and is typically ~11 AMU in winter 1963-1964 and
~14 AMU in summer near noon. Latitudinal plots of $\bar{m}_i$ show low
values over the equator increasing with latitude, markedly at night.
Above dip latitudes of about 65°, $\bar{m}_i$ ~ 16 AMU, indicating the predominance of the O^{+} ion. Thus near 1000 km, high latitude plasma
scale height values are good indicators of high latitude temperatures. These results are in good agreement with independent
measurements of the mean ionic mass in the topside ionosphere.

They emphasize the importance of the role of thermal expansion and contraction in determining the main features of the diurnal changes in the upper ionosphere. Whereas T can change by a factor of ~4 at most from one extreme to another, $\overline{m}_i$ can change by a factor of 16. Thus, except at high latitudes, the scale height, H, increases to large values at night mainly due to the decrease in mean ionic mass at the given level. At high latutudes, near 1000 km, neither $\overline{m}_i$ nor T changes very much diurnally and the scale height is constant. The latitudinal temperature profile (Fig. 3) determines the mean ionic mass distribution and, except within ± 25° dip latitude of the equator, their ratio directly dertermines the observed scale height.

In conclusion, it should be mentioned again that, up to the plasmapause, the diffusive equilbrium theory provides a reasonable explanation of the observed behaviour of the topside plasma. In particular, it accounts for the large trans-equatorial minimum in the observed scale heights, for the light-ion belt occurring over magnetic equatorial regions, for the increase in mean ionic mass, with latitude (providing a heavy ion polar ionosphere) and for the diurnal change of $\overline{m}_i$ from low values at night at a given level, to high values during the day.

IONIC RELATIVE ABUNDANCES AT 1000 km
LATITUDINAL VARIATIONS

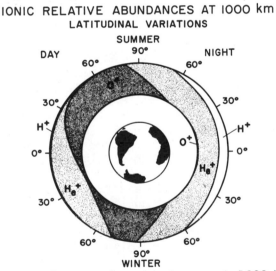

Fig. 5   Exospheric ionic relative abundances at 1000 km for different latitudes for typical day-time and night-time conditions in summer and winter in a sunspot minimum solar epoch. The relative abundance of O^+, He^+ and H^+ ions at any latitude is given by the radial width of the shaded area corresponding to each ion. The picture presented is a composite one consistent with the experimental results obtained from Alouette I, Ariel I and from whistlers (from Thomas and Dufour (1965)).

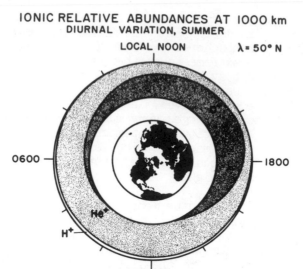

Fig. 6   Diurnal variation in the summer exospheric ionic composi-
         tion at 1000 km, at 50°N geomagnetic latitude, during a
         sunspot minimum solar epoch. The picture presented is a
         composite one consistent with the experimental results
         obtained from Alouette I, Ariel I and from whistlers (from
         Thomas and Dufour (1965)).

        This behaviour is consistent with the predictions of the dif-
fusive equilibrium theory for the basic plasma morphology and can
be summed up by the circular diagrams (Figs. 5 and 6 of Thomas and
Dufour (1965)), reproduced as Figs. 5 and 6. These figures show
the proton distributions resulting from the calculations (ibid)
combined with the relative abundances of O^+ and He^+ from the Ariel I
satellite data to give a complete picture of the ion relative abundances at 1,000 km. In deriving these diagrams, essentially the
absolute proton abundance required to account for the observed
whistler electron density profile was first computed. Once the
latitudinal dependence of this quantity had been determined, it was
assumed that the remaining O^+ and He^+ relative abundances were as
given by the Ariel I observations (Figs. 71 and 70, Bowen et al.,
1964) for the particular time and solar epoch concerned. In Figs.
5 and 6, the relative abundance of O^+, He^+ and H^+ at any latitude
is given by the radial width of the shaded area corresponding to
each ion. It should be noted, however, that in Fig. 5 the number
of O^+ ions in the night-time and the number of H^+ ions in the daytime are too small to be visible on the scale shown. This is
illustrated by a slight thickening of the lines over the regions
concerned. The theory emphasizes the basic role of temperature
changes in determining the characteristics of the plasma - a point
which might well be borne in mind as requiring further examination
in studies of the earth's plasma envelope during solar eclipses.

References

Angerami, J. J., and J. O. Thomas, Studies of planetary atmos-
      pheres, 1, The distribution of electrons in the earth's
      exosphere, J. Geophys. Res., 69, 4737 (1964).

Bowen, P. J., R. L. F. Boyd, C. L. Henderson and A. P. Willmore,
      Measurement of electron temperature and concentration from
      a spacecraft, Proc. Roy. Soc. A281, 514 (1964).

Chandra, S. and R. A. Goldberg, Geomagnetic control of diffusion
      in the upper atmosphere, J. Geophys. Res., 69, 3187 (1964).

Evans, J. V., Ionospheric backscatter observations at Millstone
      Hill, Lincoln Laboratory Technical Report 374, (1965).

Eviator, A., A. M. Lenchek and S. F. Singer, Physics of Fluids, 7,
      1775 (1964).

Goldberg, R. A., The effect of variable electron temperature on
      the equatorial electron density distribution in the upper
      ionosphere, J. Geophys. Res., 70, 655 (1965).

Thomas, J. O., and S. W. Dufour, Electron density in the whistler
      medium, Nature, 206, 567 (1965).

Thomas, J. O., and M. J. Rycroft, The exospheric plasma during
      the International Years of the Quiet Sun, J. Planetary
      and Space Sciences, in press (1970).

Thomas, J. O., M. J. Rycroft, L. Colin and K.L. Chan, The topside
      ionosphere, II, Experimental results from the Alouette I
      satellite, p322, Proc. NATO A.S.I., Finse, Norway, Editor
      J. Frihagen, North-Holland Pub. Co., Amsterdam (1966).

D-REGION CHEMISTRY

THE CHEMISTRY OF THE D-REGION

A. P. Mitra

National Physical Laboratory, New Delhi-12, India

ABSTRACT

The presentation is concerned with the following areas :
(a) the chemistries of the major neutral minor constituents :
O, O_3, H_2O and NO, (b) the negative ion chemistry in the mesosphere, (c) the individual chemistries of the positive ions NO^+,
O_2^+ and the hydrated protons. Major anomalies are identified
and discussed. Possible eclipse time changes in these parameters are also outlined.

1. INTRODUCTION

Positive ions detected and measured in the D region include
O_2^+, NO^+, metal ions, and hydrated protons, mainly H_3O^+ and $H_5O_2^+$.
Negative ions detected in the atmosphere and noticed in labora-
tory experiments include : O_2^-, O_3^-, O^-, NO_2^-, NO_3^-, CO_4^-, $O_2^- (H_2O)_n$.
Positive ion-neutral reactions proceed along decreasing ioniza-
tion potential and negative ion-neutral reactions along increasing electron affinity. During an eclipse the ion composition
in the D region remains virtually unchanged (Narcisi, Bailey
and Della Lucca, private communication), and there is no significant change in the drastic transition in ionospheric composition near 83 Km marked by the disappearance of the water cluster
ions. Severe changes in negative ion composition and concentration are expected as mesospheric atomic oxygen decreases- observations of λ (negative ion to electron ratio) by Hale et al
(this volume) give an evidence of an increase in λ by a factor
of about 40 at 70 Km.

Any aeronomic discussion of neutral and ion kinetics must depend on the use of the distributions of the neutral atmospheric constituents and on the fluxes of the relevant ionising radiations and on their photoionization yields. For O_2 and N_2 we have, in general, used the CIRA 1965 distributions.

In the mesosphere the minor neutral constituents play a very crucial role. Those known to be important include : O, O_3, NO, H_2O, CO_2 and O_2 ($^1\Delta_g$). The meagre observational information that exists and the results of photochemical considerations are discussed here in some detail. During eclipses both O and O_3 will change in the mesosphere, O possibly by a larger factor than O_3, with serious consequences in negative ion situation. No change is expected in NO, but $O_2(^1\Delta_g)$ may undergo a rapid decrease.

Values of reaction rates are, of course, crucial. Up-to-date summaries of reactions of aeronomic interest have been given for the D region by Mitra (1968) and for the entire ionosphere in the IAGA Symposium on Reactions of Aeronomic Interest held in September, 1968. The dependence of these reactions on temperature is not always known, although in some cases laboratory information has recently become available for specific ranges in temperature. The reaction rates (preferred values where the different laboratory estimates differ) are specified in the schematic charts for ions and minor particles.

2.   MINOR CONSTITUENTS

No direct observational information exists for O, H_2O and CO_2 for the mesospheric heights, with the exception of the abnormally large concentrations reported recently by Russian scientists (eg. Perov and Fedynsky, 1968). For CO_2, the usually accepted mixing ratio of 3×10^{-4} is also used here, but its uncertainty needs to be emphasised.

Ozone measurements using ultraviolet absorption are essentially limited to a height of about 70 Km (75 Km at best) above which reported values must be considered suspect. Several measurements are now available, conducted mostly with rocket-borne instruments giving the ozone distribution below this height during day as well as at night. Evans et al (1968) point out that the measurements by Johnson et al (1952), often considered the most satisfactory measurement, gives an excellent fit, between 40-70 Km, to the expression :

$$n(O_3) \;=\; 8.0 \times 10^{10} \exp\left[-(h-50)/4.53\right] \text{ molecules/cm}^3 \qquad (1)$$

Above this height, a promising technique appears to be the observation of the dayglow emission at 1.27μ from $O_2(^1\Delta_g)$. $O_2(^1\Delta_g)$ is produced primarily by photolysis of ozone (at least upto 90 Km) and has been found to be quite abundant in the mesosphere (Evans et al, 1968). At these heights quenching of $O_2(^1\Delta_g)$ is negligible and the concentration of $O_2(^1\Delta_g)$ is just the product of the production rate and the radiative lifetime. On this consideration, Evans et al (1968) obtain the following important equation :

$$n(O_2, {}^1\Delta_g) = 34\, n(O_3) \tag{2}$$

using a radiative lifetime of 3600 sec. Ozone concentration derived from Eq.(2) is larger than that would be obtained by extrapolating Eq.(1), and may well be a result of increased atomic oxygen concentration because of downward diffusion of O (Colegrove et al, 1965).

The principal reactions involved in the photochemistry of atomic oxygen and ozone alongwith their rate coefficients are shown in Fig.1. The rates are in many cases quite uncertain, and in some cases very strongly dependent on temperature. Variations of the reaction rates with latitude and seasons , of the dissociation rates with solar zenith angle, and of atmospheric density should be taken into account if the computations are to be complete.

Hesstvedt identifies three separate regions in the mesosphere : (a) 100–115 Km (b) 70–100 Km (c) 45–70 Km. The principal characteristics of these three regions are as follows :

Region I : h 100 – 115 Km

Constituent	Characteristic Time	Photochemical Status
O_3	10^2 sec	P.E.*
O	10^5 sec	
OH, HO_2	1 sec	P.E
H_2O_2	1/2 – 3 hrs	P.E.
H_2, H_2O	large	

* P.E. stands for Photochemical Equilibrium.

In this region, ozone concentration (under photochemical

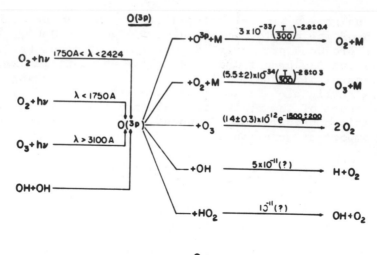

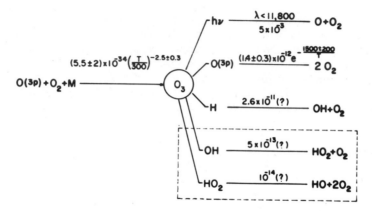

Fig. 1   Schematic diagrams on the photochemistry of O and O_3.

equilibrium)   is given by :

$$n(O_3) = \frac{5.5 \times 10^{-23}\left(\frac{300}{T}\right)^{2.5} n(O)n(O_2)n(M)}{5 \times 10^8 + 0.14 \exp\left(-\frac{1500}{T}\right) n(O) + 2.6n(H)} \qquad (3)$$

in which the term involving H may be the most dominant loss process for O_3 for this region.

Region II :   h ∼ 45 – 70 Km

All except H_2 and H_2O, have small characteristic times. $\mathcal{T}(O_3)$ is large at night and, consequently, there is no diurnal

variation. For atomic oxygen, the characteristic time at night
is about 1/2 hour, and the photochemical equilibrium value for
atomic oxygen is given by :

$$n(O)_n = 5 \times 10^{21} \left(\frac{T}{300}\right)^{2.5} \frac{n^2(OH)_n}{n(M)n(O_2)} \tag{4}$$

The day time value for atomic oxygen for heights below 58 Km,
where the characteristic time is shorter for atomic oxygen
than for ozone, is given by :

$$n(O)_d = \frac{10^{11} \; J_3 n(O_3) \; + \; 2J_2 n(O_2)}{5.5 \times 10^{-23} \left(\frac{T}{300}\right)^{-2.5} n(M)n(O_2) + 5n(OH)_d + n(HO_2)_d} \tag{5}$$

Region III :   70 - 100 Km

The characteristic time for O is increasing very rapidly
in this height range, and a marked change occurs at 80 Km,
where γ (O) is of the order of a day. O_3, OH, HO_2 and H_2O_2
have short life-times, and the daytime concentration of O_3 is
given by :

$$n(O_3)_d = \frac{5.5 \times 10^{-34} \left(\frac{T}{300}\right)^{2.5} n(M)n(O_2)n(O)_d}{10^{11} J_3 + 2.6 n(H)_d} \tag{6}$$

For H_2O and H_2 the characteristic times are large. For water
vapour it is 3 days at 100 Km and more than 1 month below
75 Km. For H_2, the time is about 1 month throughout this region.

The distributions are substantially altered when the effects
of vertical eddy diffusion and vertical mean motion are included.
The value of the vertical eddy diffusion coefficient and its
variation with height is not well known. Hesstvedt (1968) used
values varying from 4 x 10^5 cm^2/s at 65 Km to 7 x 10^6 cm^2/s at
100 Km. Shimazaki and Laird (private communication) have
tried a number of D_{eddy} models. These include constant values
of 10^6 and 10^7 cm^2 sec^{-1} and a model in which the coefficient
varies linearly from 5 x 10^5 cm^2 sec^{-1} at 40 Km to 5 x 10^6 cm^2/
sec^{-1} at 100 Km and remaining constant (at 5 x 10^6 cm^2/sec)
above 100 Km.

Hesstvedt's 'eddy diffusion model' extends from 65 to
100 Km; at these two boundaries photochemical equilibrium
values were taken. The assumption of photochemical equilibrium
is quite justified at 65 Km; but that at 100 Km is questionable,

and it would seem more realistic to start from a level (such as 120 Km) where measurements of the ratio O/O_2 exist. Calculations by Shimazaki and Laird extend from 40 to 150 Km.

The inclusion of eddy diffusion in the calculation brings in some very major changes above 80 Km, especially in the hydrogen components H, H_2 and H_2O. The concentration of water vapour which has a long characteristic time is considerably increased, taking up 85 per cent of all available hydrogen at 80 Km. There are also increases in OH and HO_2 which appear to have a serious consequence in O-distribution, since these increased concentrations now provide a very effective sink for atomic oxygen through the reactions :

$$OH \;+\; O(^3P) \longrightarrow H + O_2 \tag{7}$$

$$HO_2 \;+\; O(^3P) \longrightarrow OH + O_2 \tag{8}$$

In consequence, there is a pocket of low atomic oxygen in Hesstvedt's model near 80 Km. The other major effect on atomic oxygen is the appearance of a broad maximum centered around 90-95 Km in place of a pronounced peak at 100 Km obtained under the condition of photochemical equilibrium.

The diurnal variation of atomic oxygen at heights where the characteristic times are short (h $<$ 75 Km) is large and at lower mesospheric levels quite drastic. At 70 Km, for example, the day time concentration of 6.0×10^9 cm^{-3} (in Hesstvedt model) reduces rapidly to less than 10^5 cm^{-3} soon after sunset. The transition level at which the diurnal variation is small is determined by the value of the characteristic time; but calculations show that this is around 75 - 80 Km.

A major minor constituent is nitric oxide. This is easily ionized and in fact contributes a large part of the D region ionization. Nitric oxide appears to be quite abundant in the mesosphere, although there is considerable dispute about its actual abundance. Rocket measurements reported by Barth (1966) and very recently by Pearce (1969) place its concentration at 70 Km between $10^8 - 10^9$ cm^{-3} and at 90 Km, between 10^7 and 10^8 cm^{-3}. Ionospheric estimates at 70 Km are about one to two orders of magnitude lower (Mitra, 1968).

The neutral nitric oxide chemistry is described in Fig.2. The key reaction for mesospheric production of NO may well come from metastable $O_2(^1\Delta_g)$, first suggested by Hunten and McElroy (1968), for which a lower limit of 3×10^{-14} cm^3/s has recently been measured. With this rate NO production from $O_2(^1\Delta_g)$ will dominate in the mesosphere. Around 100 Km, however, the NO

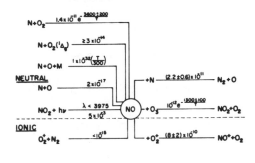

NEUTRAL NITRIC OXIDE CHEMISTRY

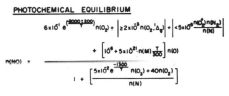

Fig.2 Schematic diagram gives the chemistry of neutral nitric Oxide.

production from O_2^+ will exceed that from any of the neutral reactions if the rate of $(O_2^+ + N_2)$ is around $10^{-16} cm^3/s$ as we now believe it to be. A large amount of O_2^+ ions are produced in the mesosphere.

$$O_2(^1\Delta_g) + h\nu \xrightarrow{\lambda\lambda 1118 - 1027 A} O_2^+ + e \qquad (9)$$

The equilibrium equation is also given in the diagram. A major sink of nitric oxide exists below about 70 Km because of its loss through O_3. The lifetime of mesospheric nitric oxide depends very critically on the amount of atomic nitrogen available, which is probably transported from the lower thermosphere (where the main source may be the dissociative recombination of NO^+). Since the rate of transfer of NO with O_2^+ is large, the quantity $n(NO^+)/n(O_2^+)$ may well be a sensitive index of the variations in the NO concentration.

For water vapour, a recent theoretical work by Hesstvedt (private communication), in which the effect of vertical eddy transport is included, gives a mixing ratio of 5×10^{-6} at 65 Km, 4×10^{-6} at 80 Km and 1×10^{-6} at 95 Km. Shimazaki and Laird give mixing rates of 1.5×10^{-6}, 1.06×10^{-6} and

0.93×10^{-6} at these heights. The chemistry of water vapour is
clearly very intimately connected with chemistry of ozone and
atomic oxygen.

Mesospheric distributions of all these minor constituents
(for day time) are shown in Fig.3. The basis on which these
distributions are obtained is given in Table 1.

3. THE NEGATIVE IONS

In the light of the current laboratory data, the negative
ion scheme takes the form indicated in Fig.4a. Uncertain
values are qualified with a question mark. The more important
channels are marked with heavier lines. Additional reactions
that may transform O_2^- to other forms of terminating ions can
be lumped together in the channel $O_2^- \rightarrow X^-$. One possible process
(and likely to be quite important) is $O^- + O_2 + M \rightarrow O_4^- + O_2$
($M=O_2$). With a 3-body rate constant of 10^{-30} cm^6/sec, it can
compete with associative detachment by O at 70 Km and below.
Another possibility is a conversion of O_2^- to hydrated negative
ions $(H_2O)_n \cdot O_2^-$.

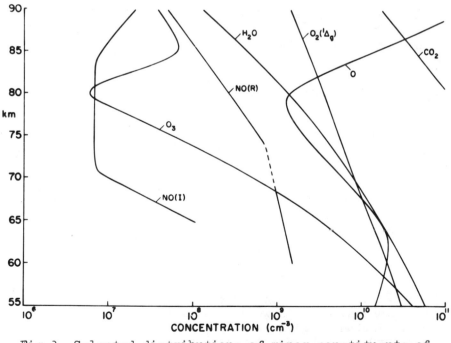

Fig.3   Selected distributions of minor constituents of
interest in D-region aeronomy.

Table 1

DISTRIBUTIONS OF MINOR CONSTITUENTS

Constituent	Source	Remarks
O	Hesstvedt (1968)	Moist atmosphere with vertical eddy transport - eddy diffusion coefficient varying from 4 x 10^5 cm^2/s at 65 Km to 7 x 10^6 cm2/s.
O_3	"	For 40-70 Km, and also approximately up to 90 Km, $n(O_3) \propto n(M)^{3/2}$.
CO_2	3 x 10^{-4} n(M)	Largest minor constituent below 85 Km for this mixing ratio. No direct observation available.
H_2O	Hesstvedt (1969)	Eddy transport included. Mixing ratios calculated to be 5 x 10^{-6} at 65 Km, 4 x 10^{-6} at 80 Km and 1 x 10^{-6} at 95 Km. With these mixing ratios, H_2O concentrations are large enough to play significant role in mesospheric ion chemistry.
$O_2(^1\Delta_g)$	Evans et al (1968)	Rocket observations of the infrared atmospheric system of oxygen in the dayglow. 50-80 Km, $n(O_2, {}^1\Delta_g) \propto n(M)^{\frac{1}{2}}$ $n(O_2, {}^1\Delta_g) = 34n(O_3)$ Above 80 Km : $n(O_2, {}^1\Delta_g) = 34n(O_3)$
NO	(a) Ionospheric Estimates: Mitra (1968 (b) Rocket measurement : Pearce (1968)	10^8 cm^{-3} at 65 Km; $\sim 10^7$ between 70-90. h>72 Km : n(NO) = 8 x 10^{-7} n(M) 60<h<69 Km : n(NO) = 8 x 10^8 e$(70-h)$

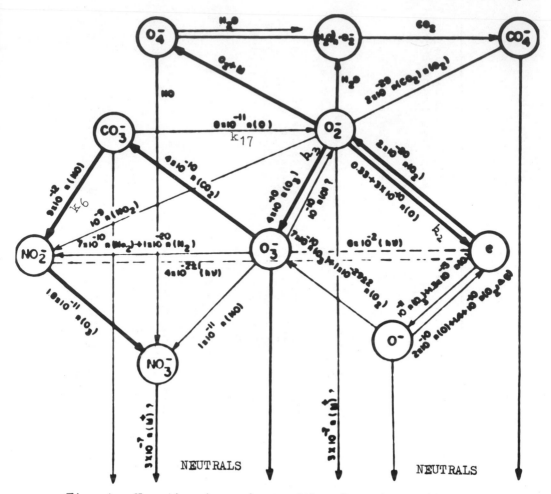

Fig. 4a   Negative ion scheme with relevant reaction rates

        An immediate consequence of this scheme is that negative
ions become very rare in the mesosphere (unless $O_2^- \rightarrow X^-$ processes dominate) in contrast with the observational estimates
(Fig.5) which indicate that λ is around unity at 70 Km during daytime.   This is the result of the very fast associative
detachment process returning O_2^- to e.   The reaction times indicated in Fig.4b for 70 Km show how fast the return step   to   e
is.   With Hunt's atomic oxygen model, λ is only 10^{-2} at 70 Km.
The situation improves somewhat if the lower oxygen atom concentrations of Hesstvedt are taken.   This allows a small portion
of O_2^- ions to charge transfer to O_3^-, and then successively
through CO_3^-, NO_2^- to NO_3^-.   The effectiveness of this diversion
depends on the link $(CO_3^- + O)$, which has to be considerably
weaker than the measured rate of 8×10^{-11} cm3/s would give

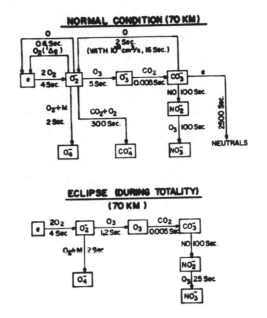

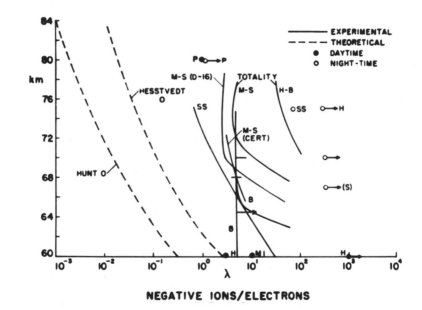

Fig. 4b   Reaction times in negative ion channels at 70 Km
          for : (a) normal times and (b) totality.

NEGATIVE IONS/ELECTRONS

Fig. 5   Observational results on negative ion to electron
         λ alongwith theoretical results (dotted lines)
         with the atomic oxygen models of Hunt and Hesstvedt.

if λ is to be as large as those indicated in Fig.5. Possible formation of ions O_4^-, CO_4^- and $(H_2O)_n$. O_2^- and their complex reactions are still essentially in the speculative stage. It has been suggested that formation of O_4^- can be sufficiently rapid to rival the associative attachment of O_2^-. The rate of the reaction $O_2^- + O_2 + M \longrightarrow O_4^- + M$ is not known, but if it is assumed to have a 3-body rate of 10^{-30} cm^6/s, its rate (per O_2^- ion) at 70 Km is 0.5 sec^{-1} against a total O_2^- detachment rate of 2.1 sec^{-1} per O_2^- ion.

It can be shown that under equilibrium conditions, the continuity equation in the case where $O_2^- \longrightarrow X^-$ is neglected is :

$$\frac{dN_e}{dt} = q - k_1 n^2(O_2)N_e \left[\frac{M}{1+M}\right] + \alpha_D n(XY^+)N_e \qquad (10)$$

where $\quad M = \dfrac{k_3 n(O_3)}{k_2 n(O) + \rho} \left[\dfrac{\alpha_i(CO_3^-)n^+ + k_6 n(NO)}{k_{17}n(O) + \alpha_i(CO_3^-)n^+ + k_6 n(NO)}\right] \qquad (11)$

Thus an attachment-type (β) loss rate would predominate in regions where $\alpha_D n(XY^+)N_e$ term is negligible, with :

$$\beta = 2 \times 10^{-30} n^2(O_2) \left[\frac{M}{1 + M}\right] \qquad (12)$$

When the terms $\alpha_i(CO_3^-)n^+$ and ρ are ignored, this expression reduces to the equations given by Bowhill elsewhere in this volume :

$$\beta = \frac{2 \times 10^{-30} n^2(O_2)N_e}{1 + R(1 + S)} \qquad (13)$$

where $\quad R = \dfrac{k_2 n(O)}{k_3 n(O_3)}$ and $S = \dfrac{k_{17}n(O)}{k_6 n(NO)}$

The crucial parameter is the relative importance of $k_{17}n(O)$ and $\alpha_i(CO_3^-)n^+$ terms. If $k_{17}n(O) \not> \alpha_i(CO_3^-)n^+$, then

$$\lambda(O_2^-) = \lambda^*(O_2^-) = \frac{k_1 n^2(O_2)}{k_2 n(O) + \rho} \qquad (14)$$

as if the $O_2^- \rightarrow O_3^- \rightarrow CO_3^-$ arm does not exist. In such a case

$$\frac{dN}{dt} = q - \alpha_D \, n(XY^+)N_e \qquad (15)$$

If, on the other hand,

$$k_{17}n(0) \not> \alpha_i(CO_3^-)n^+, \text{ then}$$

$$\lambda(O_2^-) = \lambda^*(O_2) \Big/ (1 + M) \qquad (16)$$

and Eq. 10 results.

With appropriate laboratory values of the reaction rates put in :

$$M = \frac{4 \, n(O_3)}{3n(0) + 3.3 \times 10^9} \left[\frac{3 \times 10^3 n^+ + 0.1 \, n(NO)}{0.8 \, n(0) + 3 \times 10^3 n^+ + 0.1 \, n(NO)} \right] \qquad (17)$$

If at 70 Km, $n(NO)$ is around 10^8 cm^{-3}, then under normal conditions, the term within parenthesis reduces to $n(NO)/8n(0)$. For most of the D region M is usually much less than unity.

4. POSITIVE IONS

Although only a few tens of kilometers thick, mesospheric positive ions show considerable complexity in the ionic composition. Below about 82 Km, water cluster ions $H^+.(H_2O)_n$ dominate. 32^+ ions seen in the rocket-borne mass spectrometers are believed to be S^+ below 86 Km and O_2^+ above this height. Metal ion layers have been observed near 95 Km, composed mainly of magnesium and iron, and about 5-10 Km in half-width, and a narrower layer at a somewhat higher level composed mainly of silicon, magnesium or iron. NO^+ and O_2^+ ions are the dominant ions from 90 Km to about 200 Km, where $n(0)$ equals $[n(NO^+) + n(O_2^+)]$. Positive ion chemistry must, therefore, be quite different below and above about 85 Km.

In Figs. 6 and 7 we give the reaction schemes for the positive ions NO^+, O_2^+, — the principal ions in the D region above 83 Km. Anomalies are immediately encountered. In the simplest region between 85 - 100 Km, where the hydrated protons are no longer important and negative ions negligible during daytime, and consequently the positive ion chemistry is simple, the total electron loss rate is given by :

NO$^+$ CHEMISTRY

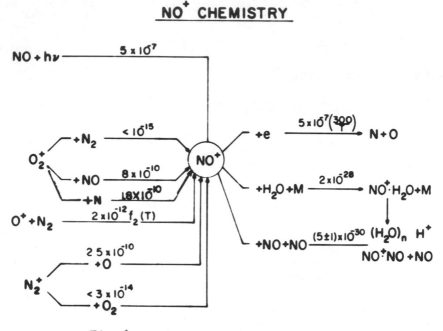

Fig. 6   Chemistry of NO+ Ions

O$_2^+$ CHEMISTRY

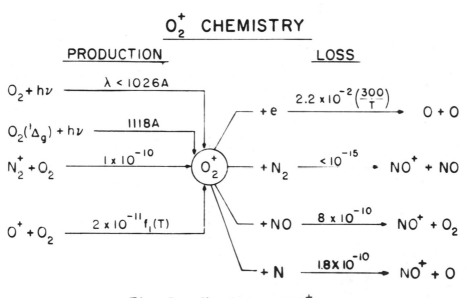

Fig. 7   Chemistry of O$_2^+$

$$\Psi = 10^{-7} \left(\frac{T}{300}\right)^{-1} \left[\frac{2n(O_2^+) + 5n(NO^+)}{N_e}\right] \tag{18}$$

giving for Ψ a value of 7×10^{-7} cm^3/s at 85 Km $(n(NO^+) \gg n(O_2^+))$ and 5×10^{-7} cm3/s at 100 Km.

For $\chi = 60°$, the measured electron density at 85 Km and 100 Km are in the neighbourhood of 2×10^3 and 3×10^4 cm^{-3} and consequently, $q(= \Psi N_e^2)$ has the following values : 3 cm^{-3}sec^{-1} at 85 Km and 5×10^2 cm-3sec^{-1} at 100 Km.

On the other hand, the production rates directly computed from the ionizing flux values are as follows :

At 85 Km and $\chi - 60°$

$q(O_2, {}^1\Delta_g) \sim 1.5$ cm^3 sec^{-1}

$q(X \text{ rays}) \sim 0.2$ " (for 0-8A X ray flux $= 2 \times 10^{-3}$ergs cm^{-2} sec^{-1})

$q(L_\alpha) \sim 3 \times 10^2$ " Pearce NO $(2.4 \times 10^8$cm$^{-3})$

$\sim 2 - 4$ " Mitra NO$(1 - 0.5 \times 10^7$cm$^{-3})$

q total $= 3 \times 10^2$ " (Pearce NO)

$= 4 - 6$ " (Mitra NO)

At 100 Km and $\chi = 60°$,

$q(O_2, {}^1\Delta_g) \sim 4$ cm^{-3}sec^{-1}

$q(X) \sim 10^2$ "

$q(L_\alpha) \sim 10^2$ " for $n(NO) = 2 \times 10^7$cm^{-3}

q total $\sim 2 \times 10^2$ "

We thus find that while at 100 Km the loss and production processes balance, at 85 Km they balance only if Mitra's low NO concentrations are used. Otherwise the available electron production rate is about two orders of magnitude larger. The complex ion production process :

$$NO^+ + H_2O + M \longrightarrow NO^+. H_2O + M \tag{19}$$

is not particularly fast at this height $(10^{-2} - 10^{-1}$/cm^3/sec).

At 70 Km, NO^+ loss rate though this process for Hesstvedt's H_2O mixing ratio of 5 x 10^{-6} is as large as 3 x $10^{-3}sec^{-1}$ per NO^+ ion, some 25 times larger than the dissociative recombination rate of 1.4 x $10^{-4}sec^{-1}$ per NO^+ ion and may even exceed the loss rate through negative ions ($\sim$ 2 x $10^{-4}sec^{-1}$ per NO^+ ion).

In the days before Hunten and McElory (1968) brought up the question of photo-ionization of $O_2(^1\Delta_g)$ the problem was to explain the sizeable O_2^+ ions, then wrongly identified in the mesosphere. Production of O_2^+ ions through X-rays is only about 10^{-1} cm^{-3} sec^{-1} at 80 Km for χ = $0^°$. To have an appreciable concentration of O_2^+ ions under these circumstances required that the process $O_2^+ + N_2 \longrightarrow NO^+ + NO$ (for which only an upper limit of 10^{-15} cm^3/s is known) must be very slow with the rate no larger than 10^{-18} cm^3/s (Mitra and Mitra, 1966). The problem has now completely reversed. O_2^+ ions are now known to be very small below 85 Km (32^+ ions previously identified are now believed to be S^+); and, in addition, O_2^+ production $O_2(^1\Delta_g)$ is copious ($\sim$ 1.6 cm^{-3} sec^{-1} at 80 Km for χ = $0^°$). If $n(O_2^+)$ is to be less than 10 cm^{-3} at 80 Km, the rate cannot be much lower than 5 x 10^{-16} cm^3/sec. This vary fast rate would have the additional effect of making the above process a major source of NO at E-region heights.

In the region below 83 Km where water cluster ions predominate, the chemistry involved is complex and as yet unresolved. Fig.8 gives a schematic diagram of a simplified version of the formation and subsequent fates of the different types of water cluster ions. Processes in which H_2O^+ is the originating step are not favoured, since its charge transfer with O_2 is quite efficient. An alternative process suggested involves a rapid production of NO^+. H_2O through process (19) which then combines further with H_2O to form H_3O^+ and with further addition of H_2O forms $H_5O_2^+$. The rate at which $H_5O_2^+$ dissciatively recombines with electrons can be crucial.

5.   CHEMISTRY OF NEUTRAL AND IONIZED PARTICLES DURING ECLIPSES

Measurements of D region electron density during solar eclipses have shown a more severe decrease in N_e than in the production rate q indicating that the loss rate must also have drastically changed. This happens at 70 Km and below. Of the neutral minor constituents that are likely to have any effect on the loss rates, the most important are O, O_3, H_2O and CO_2. Of these, changes in O and O_3 concentrations are likely to be the most important, especially those in O which are more drastic.

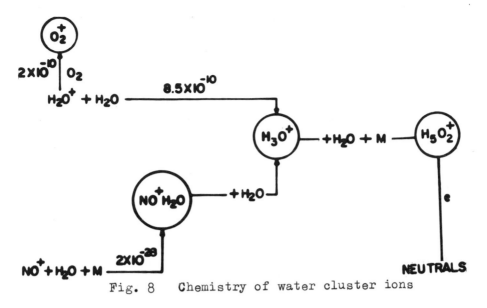

Fig. 8   Chemistry of water cluster ions

Photochemical calculations of the variations of ozone concentration in the mesosphere were carried out by Hunt (1965) for an eclipse in which the time from first contact to fourth contact is 9000 seconds and no solar radiation reaches the earth at the time of totality for 4500 seconds. Computations have also been made for two specific cases : for the eclipse of July 1963 by Doherty (private communication) and the eclipse of November 12, 1966 by Keneshea (private communication). Keneshea's recent computations consider not only the neutral atmospheric species, but also the positive and negative ions, and are thus more complete. Keneshea's computations give a few times to an order of magnitude increase in O_3 concentration between 60 – 100 Km during eclipse totality. Maximum enhancement occurs about 10 minutes after totality. Randhawa's (1968) observations with rocketborne ozonesonde during the eclipse of November 12, 1966 at Tartagal ($22^\circ 32'S$, $63^\circ 50'W$), Argentina, showed that ozone concentration at 57 Km during total solar eclipse was about 2.5 times higher than that measured a day earlier at the same altitude. The increase in ozone during total solar eclipse was found to be rapid. Various estimates of changes in O and O_3 concentrations at 70 Km are summarised in Table 2.

A large reduction in n(O) has the effect of producing a larger percentage of long enduring ions (eg. NO_3^-) by inhibiting the return channel $O_2^- + O \longrightarrow O_3 + e$ of the electrons that initially attach to O_2 to form O_2^- ions, and also by weakening the reaction $CO_3^- + O \longrightarrow CO_2 + O_2^-$. An increase in O_3 aids in this diversion. Negative ion channels during eclipsetime are shown in Fig. 4b for 70 Km.

Table 2 :   Different estimates of changes in O and O_3 concentrations at 70 Km.

	O		O_3	
	Normal	Eclipse	Normal	Eclipse
Doherty	5×10^{10}	2×10^{10}	5.7×10^9	1.5×10^{10}
Keneshea	2.3×10^9	5×10^8	1.7×10^8	5×10^8
Hunt	1.5×10^{11}	1×10^{11}	2.5×10^9	2.0×10^{10}

Some rough ideas of the changes in negative ions that will occur from changes in O and O_3 can be given. We use Eq. (17); assume that $n(NO)$ is 10^8 cm^{-3} at 70 Km, and that it does not vary during the eclipse, and take n^+ for Hale et al. For the Doherty $O-O_3$ eclipse model β increases by a factor of 12. In the Keneshea case, there is a decrease of the normally predominant O_2^- ion, and a very large increase in CO_3^- and NO_2^- ions so that the total negative ions increased by a factor of about 3 and N_e decreased by 25. The crucial role played by atomic oxygen is shown in Fig.9, in which the preeclipse values of

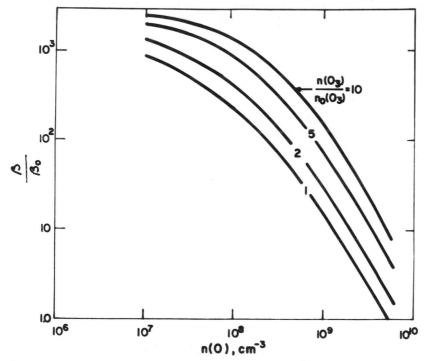

Fig.9   Increase in D region loss rate β during totality for different assumed values of O and O_3 concentrations.

$n(O)$ and $n(O_3)$ used are those given by Hesstvedt.

There is little doubt that a substantial change in λ and hence of the effective loss rate is possible during the eclipse from changes in O and O_3 concentration. The observation of a reduction in N_e at 70 Km by a factor of 500 at a time when q has been reduced to 10 per cent of its normal value requires a change in β by a factor of 50 which is only slightly larger than that observed by Hale et al, and is theoretically consistent (Fig.9) with a decrease of O to $5 \times 10^8 - 1 \times 10^9$ cm^{-3}.

No change is expected in H_2O concentration, as its characteristic time is more than one month below 75 Km and is still 3 days at 100 Km. Thus in ionic processes where H_2O may play any part (such as the production of water cluster ions or negative ions of the type $(H_2O)_n \cdot O_2^-$), there will be no significant change during the eclipse.

The other important minor constituent where any change can be crucial for mesospheric ionization is the concentration of nitric oxide. Here again, the lifetime is likely to be larger than the eclipse period and no change in NO concentration is expected, although both NO production rate if $O_2(^1\Delta_g)$ is a major participant as well as the loss rate in the lower mesosphere $(NO + O_3 \longrightarrow NO_2 + O_2)$ could be severely affected.

Substantial changes can, however, occur in the concentration of the other minor constituent responsible for mesospheric ionization : $O_2(^1\Delta_g)$. The metastables survive for their radiative lifetimes of about 45 minutes above 70 Km, and must have, therefore, nearly completely vanished at the time of the totality in both Greece and Brazil eclipses, if photolysis is the major mechanism responsible for the large $O_2(^1\Delta_g)$ densities observed at normal times. This is probably true upto 90 Km. Thus upto at least 90 Km in the D region, where photoionization of $O_2(^1\Delta_g)$ is, according to Hunten and McElroy (1968), the major source for the O_2^+ ion, there is a twofold decrease in O_2^+ concentration as the eclipse progresses; one coming from the gradual obscuration of the ionizing EUL flux and the other from the gradual disappearance of the major ionizable constituent $O_2(^1\Delta_g)$.

REFERENCES

Barth, C.A.	1966	Ann. Geophys., <u>22</u>, 198
Colegrove, F.D., W,B. Hanson and F.S. Johnson	1965	J. Geophys. Res., <u>70</u>, 4931
Delgreco, F.P. and	1968	Symposim on Physics and

and J.P. Kennealy   1968   Chemistry of the Upper
                                         Atmosphere, Waltham, Mass.,
                                         12-14 June

Doherty, R.H.   Private Communication

Evans, W.F.J., D.M.   1968   J. Geophys. Res., 73, 2885
Hunten, E.J. Llewellyn
and A. Vallance Jones

Johnson, F.S., J.D.   1954   Rocket Exploration of the
Purcell and R. Tousey   Upper Atmosphere, Ed. R. Boyd
                                         and M.J. Seaton, p. 189
                                         Pergamon Press.

Hale, L.S. and   Private Communication
Baker D.

Hesstvedt, E.   1968   Geophysical Publications of
                                         the Norwegian Academy of
                                         Sciences, 27, No.4, April

Hunten, D.M. and   1968   J. Geophys. Res., 73, 2421
M.B. McElroy

Keneshea, T.J.   Private Communication

Mitra, A.P.   1968a J. Atmosph. Terr. Phys.,
                                         30, 1065

Mitra, A.P.   1968b Third Aeronomy Conference,
                                         Illinois.

Mitra, A.P. and   1966   J. Inst. Telecomm. Engr.,
N.R. Mitra   12, 227

Narcisi, R.S., A.D.   Private Communication
Bailey and L. Della Luca

Pearce, J.B.   1969   J. Geophys. Res., 74, 853

Perov, S.P. and A.V.   1967   Xth COSPAR Assembly, London
Fedynsky

Randhawa, J.S.   1968   J. Geophys. Res., 73, 494

Shimazaki, T. and   Private Communication.
A.R. Laird

MESURES D'OZONE PENDANT LES ECLIPSES SOLAIRES

A.Vassy

Laboratoire de Physique de l'Atmosphere

Faculté des Sciences de Paris

On s'est inquiété depuis longtemps dejà des variations que pouvait subir l'ozone atmosphérique au cours d'une éclipse. Les calculs théoriques anciens laissaient prévoir en effet une augmentation de la quantité d'ozone, phénomène analogue à l'augmentation nocturne également prévue par les calculs.

Nous examinerons ce problème plus loin, à la lumière des théories les plus récentes. Pour commencer, nous allons dresser un tableau des observations faites jusqu'à ce jour et des résultats qu'elles ont apportés.

Nous avons relevé dans la littérature dix cas d'éclipses ayant fait l'objet de mesures d'ozone. La liste est portée dans le tableau suivant.

Toutes ces éclipses sauf trois ont été observées depuis le sol. Celle de 1961 a été étudiée non seulement du sol, mais encore à bord d'un avion (9, 10) et au moyen d'une fusée lancée dans la zone d'ombre (8). Nos mesures de 1966 ont été faites au moyen de radiosondes ozone ; les mesures de RANDHAWA ont été faites à bord d'une fusée traversant la zone d'ombre.

Avant d'étudier les résultats obtenus, il convient de faire quelques remarques générales sur les mesures faites depuis le sol.

1° - La variation d'ozone attendue, si elle est photochimique, ne peut concerner que les altitudes où l'ozone se forme et se détruit en moins d'une heure, soit comme nous le verrons plus loin au-dessus de 45 km. La fraction de l'ozone total au-dessus de

TABLEAU

DATE	LIEU	NATURE	OBSERVATIONS
19/06/36	JAPON	totale	I. KAWABATA (1)
9/07/45	SUEDE	totale	N. JERLOV, H. OLSSON et W. SCHUEPP (2)
25/02/52	U.R.S.S.	totale	S.A. BEZVERKNIY, A.L. OSHEROVICH et S.E.RODIONOV (3)
30/06/54	U.R.S.S. PAKISTAN SUEDE	totale	S.A. BEZVERNHIY, A.L. OSHEROVICH et S.E.RODIONOV (3) E.M. FOURNIER d'ALBE et I. RASOOL (4) B. SVENSSON (5)
19/04/58	JAPON	partielle max 0,876	M. YAMASAKI et I. SAKAI (6)
2/10/59	CONGO BELGE	partielle max 0,876	D. STRANZ (7)
15/02/61	U.R.S.S.	totale	A. L'VOVA, A. MIKIROV et W. POLOSKOV (8) R.S. STEBLOVA (9) A.K. KHRGIAN et G. KOUZNETSOV (10) G.P. GUSHIN (11)
20/07/63	U.R.S.S.	totale	Y.F. VERLAYNEN, A.L. OSHEROVICH, K.D. SUSLOV et N.S. SHPAKOV (12)
20/05/66	GRECE	partielle max 0,9996	A. VASSY et coll. non publié
12/11/66	ARGENTINE	totale	J.S. RANDHAWA (13)

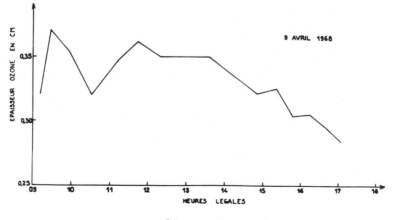

Figure I

Exemple de variations rapides de l'ozone total

45 km est très petite : très approximativement il n'y a plus que 10 à 15 unités DOBSON (U.D.) soit 1/20e de l'épaisseur réduite. La mesure doit donc s'entourer de précautions relatives à la précision, et d'autant plus que le rayonnement solaire devenant nul, les erreurs instrumentales prennent une importance relative plus grande. Un calcul effectué par FOURNIER d'ALBE et RASOOL (4) indique que l'augmentation d'ozone attendue est au maximum de 13 U.D. ; le même calcul sur des bases plus récentes fait par HUNT donne 1.6 U.D. (14).

     2° - L'épaisseur réduite subit des variations avec le temps qui peuvent être très gênantes. Si l'on a affaire à une variation régulière d'un jour à l'autre, comme cela s'observe lorsque l'on se trouve au milieu d'un anticyclone, il est facile par des mesures faites les jours précédent et suivant l'éclipse de tenir compte de cette lente dégénerescence de l'air au-dessus de la station. Mais le plus souvent on a affaire à des variations plus rapides de l'ordre de l'heure (voir un exemple fig. 1), au gré des vents circulant dans la stratosphère où se trouve la plus grande partie de l'ozone. Ces vents peuvent d'ailleurs prendre naissance du fait même de l'éclipse. Quant on assiste à une éclipse, on peut sentir le vent se lever à l'approche du deuxième contact. Ces changements dans la stratosphère ont sûrement des changements corrélatifs dans la mésosphère.

     3° - Les mesures de nature spectrographique, toutes sauf (13), doivent être corrigées de l'effet d'assombrissement du bord du disque solaire, qui est sélectif et d'autant plus important que la longueur d'onde est plus courte ; il en résulte, au cours de l'avance du disque lunaire devant le soleil, une variation de la répartition spectrale énergétique de la source qui se traduit dans l'instrument par une variation apparente de l'épaisseur d'ozone. D'après STEBLOVA, cet accroissement fictif serait de 38 U.D. pour la paire de longueurs d'onde 3112-3324 A.

4° - Une autre correction, la correction de température peut être nécessaire pour des instruments dont le pouvoir de résolution spectrale est élevé. En effet, la température influe sur certains coefficients d'absorption de l'ozone (15) et pendant l'éclipse, les températures de l'ozone varient fortement ; STEBLOVA (9) a mesuré les températures de l'ozone par la méthode de E. VASSY (16) et a trouvé une diminution de température de 60° C qui débute peu avant le 2e contact ; le retour à la normale ne s'effectue que 1 heure après le 4e contact. L'erreur   due à ce dernier effet, facile à corriger, n'est que de quelques U.D.

Nous voyons donc que l'ensemble des corrections systématiques est supérieur à la variation cherchée et les variations météorologiques également.

En gardant ces remarques en mémoire, nous allons examiner rapidement les résultats fournis par les observations depuis le sol; certains auteurs ont trouvé soit une variation beaucoup plus importante que celle attendue, soit une diminution ; il paraît clair que les variations ne sont pas dues à une modification de l'équilibre photochimique. Ainsi KAWABATA a trouvé une forte diminution de la - quantité d'ozone à partir du 1er contact suivie d'un accroissement brusque après l'éclipse ; JERLOV, OLSSON et SCHUEPP ont trouvé une baisse de 100 U.D. entre le 1er contact et le 4e contact et ont mesuré en même temps la vapeur d'eau, qui ne paraît pas avoir varié pendant cette éclipse ; BEZVERKNIY et coll. ont aussi observé une augmentation brusque après la phase de totalité. Ces observations ont dû être perturbées, comme l'ont d'ailleurs remarqué les auteurs, par les variations de nature météorologique auxquelles nous faisions allusion plus haut.

En 1954, bénéficiant d'une période de 3 jours où l'ozone restait constant, RASOOL et FOURNIER d'ALBE (4) ont noté, après la totalité une augmentation de 5 U.D. de courte durée ; SVENSSON, après corrections d'assombrissement du bord, arrive à une augmentation après la totalité de 3 à 10 U.D. suivant la longueur d'onde ; après un essai de correction de brume, il reste une augmentation de 2 à 4 U.D.

En 1958, une faible augmentation fut observée au moment du maximum, uniquement dans la station située dans l'aire d'éclipse maximale ; elle est de l'ordre de grandeur des variations sporadiques ayant précédé l'éclipse.

En 1959, une augmentation de 10 U.D. a été observée, mais elle paraît persister après la totalité.

Pour l'éclipse de 1961, les mesures en avion, à 2.500 m d'altitude, ont montré au moment de la totalité une augmentation de 30 % ; celles au sol, en différentes stations d'U.R.S.S., une augmentation à la totalité de 65 U.D. Cependant, tenant compte des variations avant et après l'éclipse, les auteurs ne pensent pas pouvoir conclure à la réalité de cette forte augmentation.

En 1963, les mesures ont été faites en plusieurs stations ce qui permet des comparaisons ; les stations hors de la ligne de totalité n'ont montré aucun effet, celles sur la ligne de totalité ont montré une croissance avant le 2e contact d'environ 3 U.D. et et une décroissance environ double après le 3e contact. On a noté également une variation des aérosols.

Comme on pouvait s'y attendre, les mesures au sol nous apportent donc surtout des éléments d'information sur les difficultés d'une mesure quantitative. Nous en retiendrons cependant que plusieurs des auteurs ont soupçonné, sans approfondir la question, que les aérosols pouvaient également perturber la mesure.

C'est dans l'espoir, non seulement d'élever l'instrument de mesure au-dessus de la partie de l'ozone qui ne subit pas de variations photochimiques rapides, mais encore d'éviter ces aérosols, que nous avons entrepris à l'éclipse de 1966 une campagne de mesures d'ozone par radiosondages ; le projet comportait un sondage pour les 2 jours précédent et le jour suivant l'éclipse à l'heure de l'éclipse, et 2 sondages pendant l'éclipse, à 1 heure d'intervalle. Ce programme a été rempli avec succès les jours sans éclipse ; mais le jour de l'éclipse, si l'opération s'est déroulée techniquement d'une façon correcte, les résultats ont causé assez de surprises pour que le dépouillement n'en soit pas encore achevé.

En effet, les observations ne peuvent s'interpréter en fonction d'une variation d'ozone, et les réponses du capteur nous semblent s'expliquer par une formation brutale d'aérosols précisèment au moment de l'éclipse totale. A ce moment, le ballon est à 14 km d'altitude. Le 2ème sondage, qui atteint cette altitude vers le dernier contact, présente une allure presque normale. Ces phénomènes étant intéressants en eux-mêmes, nous nous efforçons de les préciser ; mais on peut déjà conclure que la durée de formation de ces aérosols étant relativement courte, les méthodes de correction usuelles pour les instruments spectrographiques ne permettent pas d'en tenir compte avec précision, car au moment même de l'éclipse, les mesures sont concentrées sur l'ozone.

Restent 2 intéressantes mesures. En 1961, l'équipe
d'U.R.S.S. (8) a réussi à lancer une nacelle de mesure éjectée d'une
fusée qui a traversé la zone d'ombre vers 80-90 km pendant son parcours, permettant la comparaison entre zones éclairées et zone som-
bre. Vers 80 km, il a été observé une quantité d'ozone de 10^{-6}
cm/km, soit $2,7.10^8$ mol/cm^3 ; cette valeur est environ le double de
la valeur observée lors de tir de fusées sans éclipse. Les zones
éclairées présentaient, au cours de ce même vol, des concentrations
plus faibles. Pour comparer, rappelons que les valeurs observées de
nuit en fusée, vers la même altitude, présentent des valeurs 10^3
fois plus grandes que celles de jour. Bien entendu, ces valeurs ne
sont atteintes qu'après environ 1 heure de nuit.

Enfin, les mesures de RANDHAWA (13), faites en fusée avec
un appareil utilisant la chimieluminescence (méthode REGENER), ont
permis de mesurer l'ozone in situ ; 2 tirs ont été faits, l'un au
moment de la totalité, l'autre la veille. La comparaison des courbes (fig. 2) montre une augmentation de l'ozone sur le parcours de
l'instrument correspondant à la période de totalité ; cette augmentation à 57 km atteint plus du double de la valeur de la veille ;
au total, l'augmentation est d'environ 20 U.D. entre 50 et 60 km
(région de l'obscurité totale) ; il ne paraît pas y avoir d'effet
dans le parcours où l'éclipse était partielle. La valeur redevient
normale en moins d'une minute.

Ce résultat est donc très voisin de celui de l'équipe
d'U.R.S.S. et il semble que l'on peut conclure à la réalité de
l'augmentation observée ainsi qu'à la quasi-instantanéité du phénomène dans les hautes altitudes ; une comparaison avec les diverses
théories est donc valable, et l'on peut aussi en tirer des conclusions sur la variation de l'ozone pendant la nuit, phénomène qui
n'a jusqu'ici pas pu être mesuré de façon satisfaisante.

THEORIE DE L'EFFET D'UNE ECLIPSE SUR L'OZONE.

Les premières théories de la formation photochimique de
l'ozone étaient limitées à une atmosphère d'oxygène, ce qui simpli-
fiait beaucoup le problème. Cependant dès 1950, BATES et NICOLET
(17) considéraient le problème dans le cas d'une atmosphère oxygène-
hydrogène, mais uniquement pour la mésosphère.

Les résultats des mesures depuis le sol aussi bien qu'en
altitude montrèrent que les calculs correspondant à une atmosphère
d'oxygène donnaient des valeurs d'ozone 5 à 10 fois plus élevées
que celles observées entre 20 et 80 km. Il fallait donc apporter des
retouches sérieuses à la théorie simplifiée, qui par ailleurs pré-
sentait des insuffisances notoires, en particulier dans le cas de
la variation saisonnière.

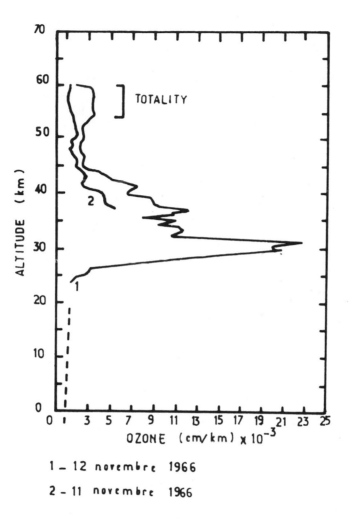

Figure 2

Résultats de RANDHAWA (13)

Courbe 1 : Au moment de l'éclipse

Courbe 2 : Le jour précédent

L'effet de la température, introduit par divers auteurs (A. et E. VASSY (18), CRAIG et G. OHRING (19)), bien qu'apportant des résultats réalistes, ne recueillit que peu de succès.

Plus récemment, HAMPSON (20) élaborait une théorie de la photochimie de l'ozone en atmosphère humide, et obtenait un meilleur accord quantitatif avec l'observation.

Puis LEOVY (21) et HUNT (22) indépendamment, au lieu de limiter les calculs au cas où l'équilibre photochimique est réalisé, parvenaient à conduire les calculs dans le cas d'un état de non équilibre photochimique, ce qui permettait d'étudier en détails la variation diurne et aussi le cas des éclipses.

HESSTVEDT (23), retouchait le modèle de HUNT en introduisant l'effet des transports verticaux entre 10 et 35 km. Et LEOVY (26) reprenait ses calculs en vue de préciser l'influence des facteurs encore incertains en les limitant à la région 15-60 km.*

Les équations régissant les réactions de la haute atmosphère sont nombreuses ; nous en donnerons la liste la plus classique, assez simple pour mieux saisir le problème, assez détaillée pour tenir compte de la complexité de cette région de l'atmosphère.

En atmosphère d'oxygène, on a :

$$(1)\ O_2 + h\nu \longrightarrow O + O \qquad \text{pour } \lambda < 2.424\ A$$
$$(2)\ O_3 + h\nu \longrightarrow O_2 + O \qquad \text{pour } \lambda < 11.800\ A$$
$$(3)\ O + O_2 + M \longrightarrow O_3 + M \qquad \text{dépend de la température}$$
$$(4)\ O + O_3 \longrightarrow 2\ O_2 \qquad \text{dépend de la température}$$
$$(5)\ O + O + M \longrightarrow O_2 + M$$

La réaction (1) conduit, suivant la longueur d'onde, à la formation de O (^{3}P) ou d'un mélange de O (^{3}P) et O (^{1}D) et pour la réaction (2) à l'un ou à l'autre de ces états de l'oxygène atomique ; aussi certains auteurs introduisent-ils une équation :

$$(6)\ O\ (^1D) + M \longrightarrow O\ (^3P) + M \qquad M \text{ pouvant être } O_2 \text{ de préférence à } N_2$$

La réaction (3) exige de l'oxygène à l'état O (^{3}P).

On voit que la formation de l'ozone n'est pas une réaction photochimique (éq. 3) ; c'est la réaction (1) donnant l'oxygène atomique nécessaire à l'équation (3) qui exige l'action de rayon-

---

* Au moment du symposium, nous avons eu communication d'une étude très complète de KENESHEA (27) ; nos conclusions n'en sont pas modifiées.

nement. Par contre, la destruction de l'ozone se fait soit par voie
photochimique directe (éq. 2), soit par collisions avec de l'oxygène atomique (éq. 4) ; mais la réaction (2) est beaucoup plus rapide
que (4) qui est négligée par certains auteurs.

Signalons au passage que l'ozone peut aussi être formé
en bombardant de l'oxygène par des électrons d'énergie supérieure
à 6,3 V ; mais cette action ne paraît pas devoir être prise en con-
sidération dans le cas d'une éclipse ; aussi nous n'en parlerons
pas.

En l'absence de lumière, il n'y a plus de formation de O
atomique, seules subsistent les réactions de disparition ; on peut
noter que la réaction (6) est supposée être très rapide et par suite O (^{1}D) disparaît pratiquement dès l'établissement de l'obscurité.
Les calculs de HUNT ont montré que les quantités d'ozone calculées
sont très sensibles à la valeur de la constante de la réaction (6)
qui est malheureusement mal connue.

HUNT considère encore la réaction :
$$(7) \quad O_2 + O_3 \longrightarrow 2 O_2 + O \ (^3P)$$

qui, en faisant disparaître l'ozone, crée de l'oxygène O (^{3}P), qui
à son tour restituera de l'ozone par la réaction (3). Aussi l'addition de cette équation (7) n'introduit que des modifications mineu-
res dans les résultats.

Toutes ces réactions n'ont pas la même importance aux diverses altitudes. On peut noter que, à la lumière, la réaction (2)
produit plus d'atomes O (^{1}D) que d'atomes O (^{3}P) au-dessus de 35 km.
Egalement, il est clair que la réaction (5) n'a que peu d'importance
dans l'équilibre au-dessousde 60 km.

Enfin, on admet que les variations de l'oxygène molécu-
laire, qui est en grand excès par rapport à O et O_3, sont sans effet
sur l'équilibre.

Revenons une fois encore sur le fait que l'équilibre photochimique n'est pas réalisé, ainsi qu'on l'a soupçonné depuis longtemps, certainement dans deux régions : au-dessous de 30-35 km, car
dans ces régions le rétablissement de l'équilibre photochimique de-
mande plus d'une journée (le tableau suivant indique les durées caractéristiques de réduction à la moitié d'une perturbation de cet
équilibre) ; et au-dessus de 60 km, parce que là c'est la concentration en atomes d'oxygène qui ne peut pas atteindre assez vite son
équilibre, et par suite celle de O_3 qui lui est liée par les équations (3) et (4). D'où l'intérêt des méthodes de calcul de HUNT et
LEOVY ne supposant pas l'équilibre réalisé.

TEMPS DE RETABLISSEMENT DE L'EQUILIBRE APRES DEVIATION DE L'EQUILIBRE

CAS DE L'OZONE		
Altitude	Réduction de l'écart à : 1/e de sa valeur HUNT (22)	Réduction de l'écart à : ½ de sa valeur LONDON (24)
20 km	13 jours	1.000 jours
30 km	22 heures	10 jours
40 km	2 heures	10 heures
50 km	17 minutes	1 heure

Il s'agit ici uniquement de l'équilibre photochimique ; si d'autres causes de déséquilibre s'introduisent (diffusion, déplacements d'air, bombardement électronique, etc...), la situation ne peut que s'écarter davantage de l'équilibre. La figure suivante due à HUNT, montre l'ordre de grandeur de l'écart entre les deux modes de calcul (fig. 3).

Toutes les réactions précédentes, compte-tenu des constantes de réaction et de leurs variations éventuelles avec la température, permettent de calculer la quantité d'ozone aux diverses altitudes et pour des conditions données de l'éclairement solaire.

Il est intéressant de noter, pour le cas des éclipses, que les modifications de concentration dues à l'apparition ou la disparition du rayonnement solaire se produisent pendant une période assez courte (moins d'une ½ heure), tout au moins au-dessus de 50 km, c'est-à-dire que ces effets seront observables même pour la faible durée d'une éclipse.

Comme cette théorie faisant intervenir seulement l'oxygène ne donnait que des résultats qualitativement satisfaisants, on a introduit l'hydrogène et ses composés, vapeur d'eau, OH, etc...

Les réactions principales (car la liste est considérable) sont les suivantes :

(8) $H + O_3 \longrightarrow OH^* + O_2$

qui conduit à l'émission des bandes de OH à partir des molécules OH excitées.

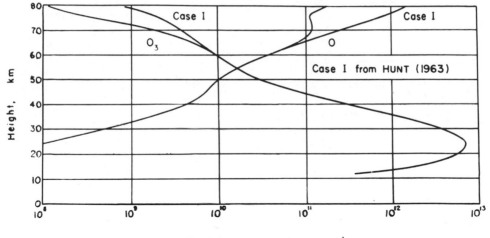

Figure 3
D'après HUNT (22)

Influence d'un état de non équilibre sur la concentration en ozone

Les molécules OH de plus bas niveau régènèrent l'hydrogène par la réaction :

$$(9) \quad OH + O \ (^3P) \quad \longrightarrow \quad H + O_2 \ .$$

Ces 2 réactions ont pour effet de détruire à la fois de l'oxygène atomique et de l'ozone, ce qui va diminuer la quantité d'ozone présente dans les régions où elle se produit ; par l'émission de OH, on peut situer entre 65 et 120 km la région où l'effet de ces 2 réactions va modifier sensiblement les résultats de la théorie à base d'oxygène seul.

Aux basses altitudes, où H a une vie très brève, cette réaction peut se remplacer par :

$$(9 \text{ bis}) \quad OH + O \quad \longrightarrow \quad HO_2 \ .$$

A cela s'ajoutent encore :

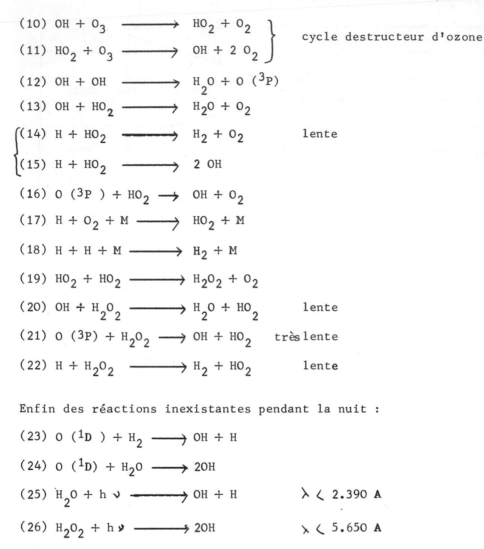

$$(10)\ OH + O_3 \longrightarrow HO_2 + O_2$$
$$(11)\ HO_2 + O_3 \longrightarrow OH + 2\ O_2$$

cycle destructeur d'ozone

$$(12)\ OH + OH \longrightarrow H_2O + O\ (^3P)$$

$$(13)\ OH + HO_2 \longrightarrow H_2O + O_2$$

$$(14)\ H + HO_2 \longrightarrow H_2 + O_2 \qquad\qquad lente$$

$$(15)\ H + HO_2 \longrightarrow 2\ OH$$

$$(16)\ O\ (^3P) + HO_2 \longrightarrow OH + O_2$$

$$(17)\ H + O_2 + M \longrightarrow HO_2 + M$$

$$(18)\ H + H + M \longrightarrow H_2 + M$$

$$(19)\ HO_2 + HO_2 \longrightarrow H_2O_2 + O_2$$

$$(20)\ OH + H_2O_2 \longrightarrow H_2O + HO_2 \qquad lente$$

$$(21)\ O\ (3P) + H_2O_2 \longrightarrow OH + HO_2 \qquad très\ lente$$

$$(22)\ H + H_2O_2 \longrightarrow H_2 + HO_2 \qquad\qquad lente$$

Enfin des réactions inexistantes pendant la nuit :

$$(23)\ O\ (^1D) + H_2 \longrightarrow OH + H$$

$$(24)\ O\ (^1D) + H_2O \longrightarrow 2OH$$

$$(25)\ H_2O + h\nu \longrightarrow OH + H \qquad\qquad \lambda < 2.390\ A$$

$$(26)\ H_2O_2 + h\nu \longrightarrow 2OH \qquad\qquad \lambda < 5.650\ A$$

Il convient de remarquer que OH, HO_2, H_2O_2 ont des concentrations toujours inférieures à celles de l'ozone, jour et nuit. Cependant NICOLET (25) considère que CH_4 est une source de OH dans la mésosphère.

Le calcul de l'équilibre résultant montre que l'introduction de réactions avec les composés hydrogénés réduit la concentration d'ozone à tous les niveaux et ramène l'épaisseur totale d'ozone à une valeur raisonnable.

Le calcul effectué dans des conditions où l'équilibre n'est pas atteint permet d'obtenir la variation diurne de l'ozone ;

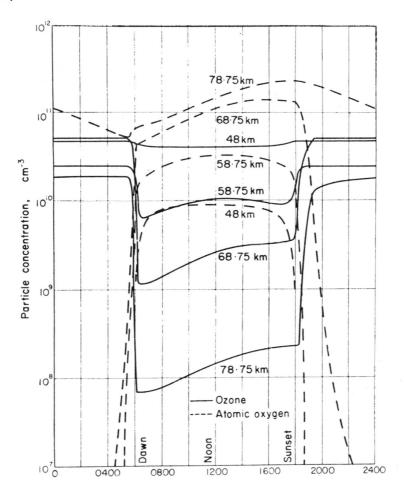

Figure 4. D'après HUNT (22)

Variation diurne de l'ozone à diverses altitudes

au coucher du soleil, c'est-à-dire dans des conditions où la disparition du soleil est brutale, l'ozone augmente dans un temps de
l'ordre de quelques minutes pour toutes les altitudes au-dessus de
50 km. Les calculs de LEOVY (21) indiquent que l'augmentation de la
concentration est de 10 fois vers 60 km, de 100 fois vers 80 km ;
ces chiffres, ont le voit, sont en bon accord avec les observations
faites en fusées lors des éclipses.

        Pour une éclipse, le calcul a été fait dans le cas d'une
atmosphère d'oxygène (14) qui donne des variations de concentration
un peut trop faibles, et dans le cas général par KENESHEA (2) qui

obtient des valeurs encore plus faibles mais encore compatibles
avec l'expérience.

     Cependant, ces théories contiennent des hypothèses diffi-
ciles à admettre, comme la constance du rapport de mélange de la
vapeur d'eau dans la stratosphère et la mésosphère. Il y a une part
d'imagination dans la liste des 26 réactions que je vous ai présentées (et j'en ai supprimé).

     D'autre part, les constantes de réaction sont souvent
mal connues et on les détermine quelquefois empiriquement de façon
à obtenir un résultat raisonnable. Il s'ensuit donc que si l'on pos-
sédait des mesures de la variation de l'ozone à différentes altitudes au cours d'une éclipse, on pourrait préciser les réactions de
la haute atmosphère, distinguer celles qui sont importantes de celles dont l'effet est mineur et voir si dans tous les éléments envi-
sagés certains ne sont pas négligeables ou inexistants. Il convient
de noter que lors d'une éclipse, la disparition du rayonnement est
progressive, de même que son retour, ce qui permet peut-être une
étude plus fine que lors du coucher ou du lever du soleil.

     Bien entendu, les mesures doivent être faites au-dessus
de 30 km

BIBLIOGRAPHIE

(1) - KAWABATA I., Jap. J. Astro. Geophys., 14, (1936) p. 37

(2) - JERLOV N., OLSSON H. et SCHUEPP W., Tellus, 6 (1954) p. 44

(3) - BEZVERKHNIY S.A., OSHEROVICH A.L. et RODIONOV S.F., Dok.
      Akad. Nau. SSSR, 106, (1956) p. 651

(4) - FOURNIER d'ALBE E.M. et RASOOL I., Ann. Géophys., 12 (1956)
      p. 72

(5) - SVENSSON B., Arkiv Geof., 2 (1958) p. 573

(6) - YAMASAKI M. et SAKAI I., Tateno Aer. Obs. Jour., 6 (1959)
      p. 97

(7) - STRANZ D., Tellus, 13 (1961) p. 276

(8) - L'VOVA A., MIKIROV A. et POLOSKOV X., Geom. Aerono (URSS), 4
      (1964), p. 1082

(9) - STEBLOVA R.S., Dok. Akad. Nauk SSSR, 2 (1962) p. 148

(10) - KHRGIAN A.K. et KOUZNETSOV G., Gidrometizdat, (1965), p. 26

(11) - GUSHIN G.P., Etudes de l'ozone atmosphérique, Gidrometizdat (1963) P. 249

(12) - VEROLAYNEN Y.F., OSHEROVICH A.L., SUSLOV K.D., et SHPAKOV N.S., Geomagn. Aerono. (URSS), 5 (1965) p. 81

(13) - RANDHAWA J.S., J. Geophys. Res., 73, 1968, p. 493

(14) - HUNT B.G., Tellus, 17, (1965) p. 516

(15) - VASSY A. et VASSY E., C.R., 128, (1949) P. 764

(16) - VASSY E., Ann. Geophys., 8, (1947), p. 679

(17) - BATES D.R. et NICOLET M., J. Geophys. Res., 55, (1950) p. 39

(18) - VASSY A. et VASSY E., Jour. Phys., 2, (1941), p. 81

(19) - CRAIG R.A. et OHRING G., Jour. Meteor., 15 (1958), p. 59

(20) - HAMPSON J., Problèmes météorologiques de la stratosphère et de la mésosphère, P.U.F., (1966), p. 393

(21) - LEOVY C., J. Atm. Sc., 21 (1964) p. 238

(22) - HUNT B.G., Atm. Terr. PHys., 27, (1965), p. 133
                J. Geophys. Res., 71, (1966), p. 1385

(23) - HESSTVEDT E., Geofysike Publ., 27, 1968 n° 5

(24) - LONDON J., Space Res. VII, (1967), p. 172 (North Holland)

(25) - NICOLET M., Geophysics the Earth's Environment, Gordon et Breach, (1963) p. 202

(26) - LEOVY C., J. Geophys. Res., 74,(1969) p. 417

(27) - KENESHEA T.J., Univ. Illinois Aeronomy Report n° 32, p. 400

DIRECT ROCKET MEASUREMENTS OF THE IONOSPHERE

DURING THE SOLAR ECLIPSE OF NOVEMBER 12,1966

L.C.Hale

Ionosphere Research Laboratory

The Pennsylvania State University,University Park,Pa.

Direct measurements of ionospheric parameters were made on a number of rockets fired in conjunction with the solar eclipse of November 12, 1966 from a site in southern Brazil. Previously reported results indicate unexpectedly large decreases in the electron density at eclipse totality in the vicinity of 80 km. altitude. Also, measurements with the blunt probe technique showed an unexpectedly large decrease in positive ion density during the course of the eclipse. This result leads to an ion-ion mutual neutralization coefficient, α_1, of order 10^{-6} cm^3/sec., and a nitric oxide density of order 10^8/cm^3 in the altitude range 65-78 km. The appropriate ion chemistry model was found to agree with that proposed by Fehsenfeld, et. al. below 60 km., but to diverge greatly from this model at higher altitudes.

An examination of possible ion chemistry models is given, in which it is shown that a general model is greatly complicated by the probable large differences in recombination coefficients for different ionic constituents. However, it is shown that a general model must reduce to a two negative ion model for some specific cases, including that of the eclipse. Under this assumption, the blunt probe data is used to demonstrate the existence of a process (at present unknown, but possibly involving atomic hydrogen) to initiate the conversion of O_2^- ions into stable negative ions. This process increases with altitude even more rapidly than that calculated by Adams and Megill, and is the dominant O_2^- loss process above about 76 km.

The blunt probe data is extended to calculate electron loss coefficients to compare with other electron density data. It is

81

shown that the electron density continuity equation in the vicinity of 80 km. reduces to a linear, first order differential equation with a time constant of less than one minute, both for eclipsed and full sun conditions.

INTRODUCTION

A number of direct measurements of the ionosphere were made in conjunction with the solar eclipse of November 12, 1966, from a site near Cassino, Rio Grande do Sul, Brazil. The measurements with which I was directly concerned, using the parachute borne blunt probe technique,[1] have been reported in a paper by Baker and Hale.[2] What follows here is a direct extension of the Baker-Hale paper, in which the results of that study are extended in such a way as to be comparable to the electron density results of other experiments.

RESULTS OF BAKER-HALE STUDY

A recapitulation of the Baker-Hale paper, "D-Region Parameters from Blunt Probe Measurements During a Solar Eclipse,"[2] follows.

Five rockets carrying parachute borne blunt probes[1] and also BRL Langmuir probes and Penn State propagation experiments were launched in conjunction with the eclipse. The blunt probes measured positive conductivity and negative conductivity of the atmosphere between about 45 and 78 km. Using a small ion model and an assumed mobility, positive ion densities were determined.

The use of an assumed mobility is not believed to be inaccurate, since mobility is a very slow function of mass (approximately the square root of the reduced mass.) In any case it was primarily the variation and not the absolute values of the ion density which was used in drawing further conclusions.

Data on electron drift velocity vs. E/P was used to convert the negative conductivity measurements to electron density profiles. These profiles compared closely with other direct measurements of electron density above 70 km., and diverged greatly from them below 65 km., giving much higher values of electron density.

The positive ion-negative ion recombination coefficient, α_i, was obtained using a computer integration of the positive ion continuity equation, using the totality and full-sun positive ion profiles as boundary conditions. The results were shown for various extreme assumptions about X-ray and galactic cosmic ray ionization sources, and it was seen that these various assumptions make little difference to the conclusion that α_i is of order 10^{-6} cm^3/sec. in the altitude range 65-78 km. It was concluded that ionization production primarily involved the ionization of nitric oxide by solar Lyman-α, and the NO density was of order 10^8/cm^3 from 65 to 78 km.

The ratio of the full-sun negative ion to electron density, λ, was seen to increase with altitude above some altitude which depends on the particular electron density profile used. This is contrary to previous expectations, and lends support to the viewpoint that a stable negative ion is formed readily in the upper D-region in the daytime.

A comparison of the data with a model incorporating the negative ion scheme of Fehsenfeld, et. al.,[3] showed good agreement with this model only below 60 km., and diverged widely at higher altitudes, with the model failing by one to two orders of magnitude to produce the observed numbers of negative ions at 80 km.

A "GENERAL" D-REGION MODEL

The principal D-region parameter deduced from the Baker-Hale[2] study is an "effective" positive ion-negative ion recombination coefficient, α_i. To consider what this quantity means, a typical "general" ion chemistry model is diagrammed in Fig. 1. The structure of this model is not meant to be definitive. For example, some workers would contend that not all ions in the positive ion "chain" originate from NO$^+$, others that parallel chains of positive and/or negative ions may be involved. What is illustrated is that, in general, α_i (and also α_d, the electron-positive ion recombination coefficient) are weighted averages of a number of different reaction coefficients. It might be expected that the more complex ions formed along the chains would have larger recombination coefficients. In a recent discussion with Professor Massey,[4] he indicated that recombination reactions involving complex ions, such as hydrated ions, would be expected to have very much higher rate coefficients than simpler ions. He said that he believes this to be primarily due to their ability to "distend" themselves in such a way as to greatly prolong the time of the interactions, rather than to the existence of more "channels" by which the reactions can proceed. I think that the complexity of the situation is such that we should be very careful about talking about such things as "effective recombination coefficients," except for very well understood specific situations.

THE TWO NEGATIVE ION MODEL

I now wish to discuss the kind of model which results if the general model discussed above, or any other "general" model,

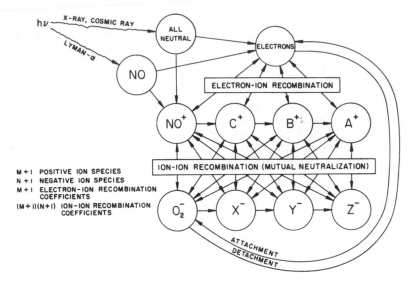

FIG. I. ION CHEMISTRY MODEL

is simplified for a specific situation. If in a particular
situation one kind of positive ion and one kind of negative ion in
addition to O_2^- are important (or, if there are more kinds of
ions, that they have the same recombination coefficients), and
if essentially all attachment and detachment is to and from the
initial negative ion O_2^-, then the "general" model reduces to the
two negative ion model shown in Fig. 2. [This model is similar
to that proposed by Adams and Megill,[5] the only difference
being the inclusion here of two different values of the ion-ion
recombination coefficient.] These conditions are certainly true
for the eclipse, where the predominant ions are certainly the
terminal ions A^+ and Z^- of Fig. 1. [They may not be true for
the twilight situation considered by Adams and Megill,[5] where
different terminal ions may exist by night and day.]

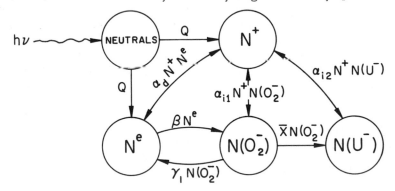

FIG. 2. TWO NEGATIVE ION MODEL
AFTER ADAMS AND MEGILL

Since it is desired to present the blunt probe eclipse data in a form which can be compared with other experiments in which electron density is the observable quantity, the electron density continuity equation for the two negative ion model, along with some other easily derived relations, are considered [Table A]. In these equations, β is the three body attachment coefficient for the formation of O_2^-, γ_1 is the O_2^- detachment coefficient (involving reactions with O and with $O_2(^1\Delta_g)$), and $\overline{X}$ is the coefficient of the reaction or reactions which initiate the conversion of O_2^- into a stable negative ion U^-. The other symbols have their standard meanings.

$$\frac{dN^e}{dt} = Q + \gamma_1 N(O_2^-) - \beta N^e - \alpha_d N^+ N^e \approx 0$$

$$N(O_2^-) = \frac{\beta}{\gamma_1 + \overline{X} + \alpha_{i1} N^+} N^e$$

$$\frac{dN^e}{dt} = Q - \frac{\beta}{N^+} \frac{1}{1 + \dfrac{\gamma_1}{\overline{X} + \alpha_{i1} N^+}} + \alpha_d N^+ N^e$$

$$\lambda = \frac{\beta}{\gamma_1 + \overline{X} + \alpha_{i1} N^+} \left(1 + \frac{\overline{X}}{N^+ \alpha_{i2}} \right) = \frac{N(O_2^-) + N(U^-)}{N^e}$$

EQUATIONS FOR TWO NEGATIVE ION MODEL

TABLE A

Using the eclipse blunt probe data and using the Q model of Hunt[6] and the $O_2(^1\Delta_g)$ model of Hunten and McElroy,[7] Baker[8] has calculated the coefficient $\overline{X}$ [Fig. 3]. This is compared with the $\overline{X}$ deduced by Adams and Megill[5] from twilight riometer data during a PCA. It is seen that both are rapidly increasing with altitude, although that deduced by Baker is an even more rapidly increasing function of altitude than that of Adams and Megill. It will also be noted here that the altitude dependence of $\overline{X}$ is absolutely wrong for it to involve ozone, which is the first step in the Fehsenfeld, et. al.[3] negative ion chain. [In fact, the only neutral species expected to possess this kind of altitude dependence is atomic hydrogen,[9] and it is suggested the concentration of H and the products and rates of the reaction between O_2^- and H should be given further study.]

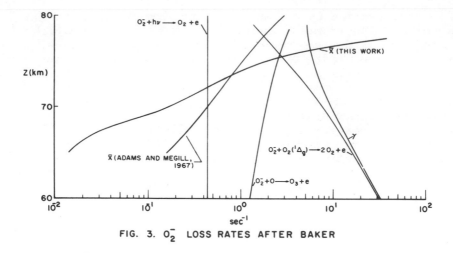

FIG. 3. O_2^- LOSS RATES AFTER BAKER

For the purposes of comparison with various electron density measurements, the quantity

$$\alpha = \alpha_d + \frac{\beta}{N^+} \cdot \frac{1}{1 + \dfrac{\gamma_1}{\overline{X} + \alpha_{i1} N^+}}$$

as determined using the blunt probe data has been calculated at various altitudes (Table B). This quantity is of the nature of an effective recombination coefficient.

DISCUSSION OF RESULTS

It will be noted that α varies during the course of the eclipse because of its dependence on N^+, and the variation of N^+. The values shown in Table B are for the full-sun conditions just after the eclipse.

Because of the uncertainty in the contribution of γ_1 due to atomic oxygen, both due to the lack of precise knowledge of the O density and the fact that it will certainly vary during the course of the eclipse, the effect on α of a variation of O from the model value to zero concentration is shown in Table B. It is seen that this effect is confined to a region between 60 and 80 km., with really substantial effects confined to a layer between 70 and 78 km., with the maximum effect being about a 55% increase in α at about 75 km.

HEIGHT (km)	$\alpha = \alpha_d \dfrac{\beta}{N^+} \dfrac{1}{1 + \dfrac{\gamma_1}{\overline{X} + \alpha_{il} N^+}}$	% VARIATION If $N(O) \to O$	REMARKS
65	1.54×10^{-6} cm^3/sec	5%	$\to \alpha_d + \dfrac{\beta \overline{X}}{N^+ \gamma_1}$
70	2.94×10^{-6}	15%	
72.5	4.08×10^{-6}	23%	
75	5.38×10^{-6}	55%	$\overline{X} = \dfrac{1}{3} \gamma_1$ $\leftarrow K(O) = K(O_2\,^1\Delta g)$
76.25	7.2×10^{-6}	30%	$\overline{X} = \gamma_1$
77.5	6.13×10^{-6}	12%	$\overline{X} = 4\gamma_1$
80 (extrapolated)	3.8×10^{-6}	1%	$\to \alpha_d + \dfrac{\beta}{N^+}$

α, "EFFECTIVE" ELECTRON-POSITIVE ION RECOMBINATION COEFFICIENT FULL SUN CONDITIONS

TABLE B

At the top of the altitude range covered, near 80 km., the electron loss is almost completely dominated by attachment, and hence the suggestion of Prof. Bowhill that we should cast our models in a form involving an effective attachment coefficient,

$$\frac{dN^e}{dt} = Q - BN^e$$

is certainly a good idea.

For this model,

$$B = \alpha_d N^+ + \beta \frac{1}{1 + \dfrac{\gamma_1}{\overline{X} + \alpha_{i1} N^+}}$$

and reduces to the forms

$$B_{65} = \alpha_d N^+ + \frac{\beta \overline{X}}{\gamma_1} \qquad \text{below 65 km.}$$

and $\qquad B_{80} = \alpha_d N^+ + \beta \qquad\qquad$ in the vicinity of 80 km.

There are approximately equal contributions from the two terms of B_{65} at 65 km., and the form of the equation shows the relative importance of recombination, attachment, detachment, and the charge transfer process $\overline{X}$. In the vicinity of 80 km., however, it is seen that the $\overline{X}$ process is so rapid as to make detachment unimportant, and the dominant term is simply β, the actual attachment coefficient. A consequence of this is that the electron density in the vicinity of 80 km. is described approximately by a linear first order differential equation. The full-sun time constant associated with this equation, $\tau_{80} = 1/B_{80}$, is found to be about 50 sec., and will of course decrease slightly from this value at totality due to a decrease in N^+. Another consequence of this first order behavior is that in the vicinity of 80 km. N^e is proportional to the electron production Q and not to $Q^{1/2}$.

The consequences of this reduction of the model to a linear first order differential equation would be expected to persist above 80 km. to the altitude at which the electron loss by recombination is of the same order as that due to attachment, which should occur at approximately 83 km.

A few words are in order about the applicability of the first order linear model derived from the eclipse to other geophysical situations. It must be pointed out that the coefficients of the equations of the two negative ion model (Table A) have been evaluated for a quasi-equilibrium situation

$$\frac{dN^e}{dt} \approx 0 .$$

Nevertheless, the equations for the loss coefficients α and B would be valid for other situations, including dynamic situations, if the basic physical situation was the same. This means that the equations should be valid for the undisturbed slowly varying daytime D-region, and for dynamic situations involving even rapid downward changes in the ionizing radiation, such as occur during a solar eclipse. Upward changes in ionizing radiation, such as X-ray flares, which may cause large numbers of ions to exist temporarily in "intermediate" states of the positive and/or negative ion chemistry chains, would naturally complicate the situation, and the simple model presented above would probably not be adequate to describe such situations correctly.

ACKNOWLEDGEMENTS

This paper was written while the author was on sabbatical leave from the Ionosphere Research Laboratory, of The Pennsylvania State University and also supported by a Fellowship from the Belgian Americal Educational Foundation.

The blunt probe data was obtained during the eclipse of November 12, 1966 on a program sponsored by DASA and the U.S. Army Ballistic Research Laboratory, Aberdeen Proving Ground, Maryland.

REFERENCES

1. Hale, L. C., D. P. Hoult and D. C. Baker, Space Research VIII, 320, North Holland, Amsterdam (1968).

2. Baker, D. C. and L. C. Hale, Paper presented at COSPAR, Prague, May 1969, in press, Space Research X, North Holland, Amsterdam.

3. Fehsenfeld, F. C., A. L. Schmeltekopf, H. I Schiff and E.
    E. Ferguson, Planetary and Space Science, 15, 373 (1967).

4. Massey, H. S. W., Private communication, Prague, May
    1969.

5. Adams, G. W. and L. R. Megill, Planetary and Space
    Science, 15, 1111 (1967).

6. Hunt, B. G., Journal of Geophysical Research, 71, 1385
    (1966).

7. Hunten, D. M. and M. B. McElroy, Journal of Geophysical
    Research 73, 2421 (1968).

8. Baker, D. C., Scientific Report No. 334, Ionosphere
    Research Laboratory, The Pennsylvania State University,
    February 1, 1969.

9. Hesstvedt, E., Geofys., Publ., Vol. 27, No. 4, April 1968.

SOLAR RADIATION AND LAYER THEORY

OBSERVATIONS OF SOLAR IONIZING RADIATIONS AND THEIR SIGNIFICANCE FOR THE ECLIPSE-IONOSPHERE

H.Friedman

U.S. Naval Research Laboratory

Washington, D.C.20390, U.S.A.

In recent years space research has provided an ever increasing body of data on the solar spectral intensity distribution and on the spatial emission pattern of ionizing radiation. From these data we can calculate in detail the time variations of the fluxes of ionizing wavelengths arriving at the ionosphere during the course of an eclipse. When such complete flux data are available simultaneously with ionospheric soundings it should be possible to make discriminating tests of ionic reaction rates and electron production and loss processes. In the following paragraphs examples are given of the kinds of spectral flux data that have been obtained and of some specific eclipse observations.

IONIZING RADIATION IN THE UPPER ATMOSPHERE

For the present discussion, we shall consider various height ranges and the radiations that contribute to ionization within those ranges. Fl may be taken as the range 140-200 km. Higher levels belong to F2. The primary sources of ionization are the same in both ranges, but the major absorption of energy is in Fl. The important ionizing rays include the Lyman continuum and the wavelength band 350-200 A, which contains about 1 erg/cm^2/sec primarily in He II (304 A). Wavelengths from 650-450 A are preferably absorbed between 160 and 180 km, and less than 10% penetrates below 145 km. Included in the 650-450 A range are the resonance lines of He I (584 A) and lines of Mg X (625, 610 A) and Si XII (520, 500A). The total flux adds up to about 0.4 erg/cm^2/sec.

From 170 A to about 211 A there is a dense grouping of lines comprising about 1 erg/cm^2/sec. These lines originate primarily in stages of Fe VIII through XIV. In contrast to Fe XV and XVI, which are concentrated over plages, the lines from 170 to 211 A emerge almost uniformly from the entire disk.

E region encompasses the altitude range 90-140 km, in which the effective radiations include x-rays (10-100 A), H Lyman β (1025.7 A), C III (977 A) and the longer wavelength portions of the Lyman continuum (910-800 A). Near solar maximum, the observed flux of Lyman β was about 0.05 erg/cm^2/sec, and of CIII, 977 A, about 0.08 erg/cm^2/sec. The x-ray spectrum may be divided into two parts, 10-31 A and 31-100 A, having roughly the same absorption cross sections. Most of the energy is contained in the longer wavelength band which originates over most of the disk. The shorter wavelengths of the x-ray spectrum are localized and highly variable.

The region 1350-1750 A does not contain any radiation capable of ionizing the major constituents of the upper atmosphere in their normal states, but this range is important to the ionosphere because it dissociates oxygen. The flux given by the U. S. Naval Research Laboratory (NRL) measurements (1954, 1960) is about 3.5 x 10^{11} protons/cm^2/sec. No information is available on variability. However, the observed flux implies a dissociation time of the order of two days and other evidence implies substantially longer times. The influence of an eclipse on the ratio O/O_2 is therefore essentially negligible.

The data of Table 1 are taken from the summary by Hinteregger[1] and are representative of the quiet sun from 1963-66.

DISTRIBUTION OF X-RAY EMISSION FROM THE SOLAR DISK

An important advance in our understanding of the spatial distribution of x-ray emission on the sun came in 1960 when the NRL group succeeded in obtaining an x-ray image (20-60 A) of the sun with a pinhole camera.[2] Active centers of x-ray emission coincided with bright plages. The x-ray emission from the brightest plage area was at least 70 times as intense as the quiet background. Evidence was also found for limb brightening, as much as a factor of 2, in spite of the relatively poor camera resolution which was about 0.1 solar diameter.

A variety of evidence relating to active centers has been derived from the series of SOLRAD satellites instrumented by NRL. In March 1966, SOLRAD observed x-rays associated with the first major spot groups of the new cycle, Fig. 1. Following a

Table I   Estimated Values of Solar XUV Fluxes for conditions of low activity
during the IQSY

	Spectral Group	Wavelength or Wavelength Range (Å)	Φ_0 (photons cm^{-2} s^{-1})
I		1775–1325	2.7×10^{12}
II	(incl. HI, λ 1215.7)	1325–1027	3.5×10^{11}
Group 1			
1.1	H I	1025.7	2.7×10^9
1.2	N III	991.5	0.4
1.3	(excl. 1.1, 1.2)	1027–990	1.5
1.4	C III	977.0	4.0
1.5	H I	972.5	0.65
1.6	(excl. 1.4, 1.5)	990–950	0.6
1.7	H I	949.7	0.35
1.8	H I	937.8	0.2
1.9	(excl. 1.7, 1.8)	950–911	1.5
Total, Group 1		1027–911	11.9×10^9
Group 2			
2.1	H Ly–cont	911–890	4.0×10^9
2.2	H Ly–cont	890–860	3.8
2.3	H Ly–cont	860–840	1.8
2.4	O II, III	832–835	0.65
2.5	(excl. 2.4)	840–810	1.6
2.6		810–796	0.7
Total, Group 2		911–796	12.6×10^9
Group 3			
3.1	O IV	790.1	0.36×10^9
3.2	O IV	787.7	0.28
3.3	Ne VIII	780.3	0.16
3.4	(excl. 3.1, 3.2, 3.3)	796–780	0.5
3.5	Ne VIII	770.4	0.32
3.6	N IV	765.1	0.23
3.7	(excl. 3.5, 3.6)	780–760	0.6
3.8		760–740	0.4
3.9		740–732	0.15
3.10	O III	703.8	0.20
3.11	(excl. 3.10)	732–700	0.35
3.12		700–665	0.7
3.13		665–630	0.35
Total, Group 3		796–630	4.6×10^9

Table I—Continued

Spectral Group		Wavelength or Wavelength Range (Å)	Φ_0 (photons cm^{-2} s^{-1})
Group 4			
4.1	O V	629.7	1.0×10^9
	Mg X	625	0.2
4.2	(excl. 4.1; but incl. Mg X, λ 610)	630–600	0.8
4.3	He I	584.3	1.0
4.4	(excl. 4.3)	600–580	0.25
4.5		580–540	0.8
4.6		540–510	0.5
4.7		510–500	0.5
4.8	(incl. Si XII, λ 499.5)	500–480	0.6
4.9	(incl. Ne VII, λ 465.3)	480–460	0.4
Total, Group 4		630–460	6.1×10^9
Group 5			
5.1		460–435	0.35×10^9
5.2		435–400	0.6
5.3		400–370	0.4
Total, Group 5		460–370	1.4×10^9
Group 6			
6.1	Mg IX	368.1	0.4×10^9
6.2	(excl. 6.1; incl. Fe XVI, λ 360.8)	370–355	0.7
6.3		355–340	0.6
6.4	(incl. Fe XVI, λ 335.3)	340–325	0.4
6.5		325–310	0.4
6.6	He II	303.8	4.0
6.7	(excl. 6.6; incl. Fe XV)	310–280	0.8
Total, Group 6		370–280	7.3×10^9
Group 7			
7.1		280–260	0.5
7.2	(incl. Si X, He II, λ 256)	256	0.3
7.3	(excl. 7.2)	260–240	0.4
7.4		240–220	0.4
7.5		220–205	0.3
Total, Group 7		280–205	1.9×10^9

Table I—Continued

	Spectral Group	Φ_0 (photons cm^{-2} s^{-1})	I_0 (erg cm^{-2} s^{-1})
Group 8			
8.1		205–190	1.6×10^9
8.2		190–180	2.2
8.3		180–165	3.2
Total, Group 8		205–165	7.0×10^9
Group 9			
9.1	165–130	4×10^8	0.052
9.2	130–120	0.05	0.0008
9.3	120–110	0.16	0.0028
9.4	110–100	0.5	0.0094
9.5	100– 90	0.8	0.016
9.6	90– 80	0.8	0.017
9.7	80– 70	0.6	0.016
9.8	70– 60	0.6	0.018
Total, Group 9	165–60	7.5×10^8	0.13
Group 10			
10.1	60– 50	0.6×10^8	0.021
10.2	50– 40	0.3	0.013
10.3	40– 31	0.3	0.017
Total, Group 10	60– 31	1.2×10^8	0.051
Group 11			
11.1	31– 22.5		< 0.002
11.2	22.5– 8		< 0.001
11.3	8– 1		< 0.0001
Total, Group 11	31– 1		< 0.003

sunspotless week, the new group appeared at the limb on 15 March.
X-rays were recorded two days before the spot came around the
limb, Fig. 2. Within a few days the 8-20 A flux exceeded 50 times
the quiet sun background from the entire disk, even though the
active region covered less than 1/1000th of the area of the disk.
The corona above the active spot group must, therefore, have
reached an x-ray brightness more than 50,000 times as great as
the surrounding corona.

Cosmos Satellites 166 and 230 of the U.S.S.R. carried x-ray
heliographs for the 2-8 A and 8-14 A bands. From observations
of the limb occultations of flares and active regions, Beigman
et al[3] were able to establish the heights of x-ray emitting
regions. Figure 3 shows the observations of the variation of
x-ray flux on 16-18 June 1967 as the region M 8836 passed behind
the limb. The solid curve gives the height, H, of the point
above the active region which is occulted as a function of the
elapsed time from when the plage 8836 was at the limb. The last
x-ray flare originated in 8836 at 04 h on 17 June and was, therefore, masked at heights below 20,000-25,000 km. This height
agrees with optical estimates of $H > 15,000$ km for limb flares
connected with sudden ionospheric disturbances (SID). It is also
evident in Fig. 3 that x-ray emission from the active region in
the absence of flares was recorded up to 01 h on 8 June, which
means that the height of emission exceeded 80,000 km.

Friedman[4] has reviewed the variability as revealed by SOLRAD
satellites as follows:

In July 1964 the solar x-ray flux reached its minimum level.
The 44-60 A photometer of SOLRAD 1965-16D recorded 0.02 erg/cm^2/sec,
close to the minimum value observed in a few scattered rocket
measurements during the previous minimum of 1953-1954. By January
1967, the level had risen to 0.07 erg/cm^2/sec and appeared to be
rapidly approaching the maximum of flux reached in 1957-1958. All
through the solar minimum period, the 44-60 A photometer gave
signals of measurable flux at the 0.02 erg/cm^2/sec level, or
greater. At 8-12 A, signals were sporadic at a minimum detectable
level of 1.0 x 10^4 erg/cm^2/sec from January to April 1964, disappeared completely in May through July, and appeared again
sporadically from August to December. During March 1965, instruments aboard OSO-2 recorded continuously detectable fluxes
(8-20 A) between 2 x 10^{-3} and 4 x 10^{-3} erg/cm^2/sec, but the flux
(0-8 A) rarely exceeded 2 x 10^{-5} erg/cm^2/sec. In the first part
of 1967 flux levels (0-3 A) observed by SOLRAD almost continuously
exceeded 2 x 10^{-6} erg/cm^2/sec; F(0-8 A) fluctuated above a base
level of 1 x 10^{-3} erg/cm^2/sec and F (8-20 A), above a base level
of 2 x 10^{-2} erg/cm^2/sec.

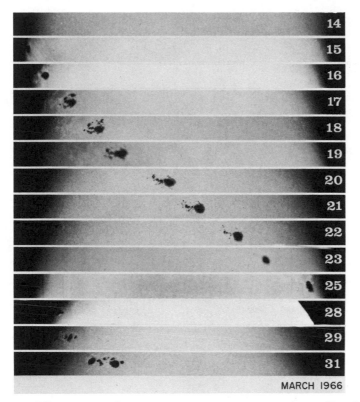

Fig. 1.   On 15 March a large spot group appeared on the limb of
the sun.   As it traveled across the disk from 15 March to 25 March
it was the only sunspot on the disk.   On 28 March a new spot group
appeared on the limb.   Figure 2 shows the x-ray flux attributable
to these spot groups.

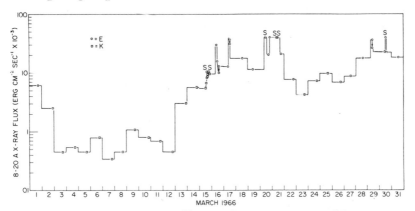

Fig. 2.   Solar x-ray fluxes (8-20 A) for March 1966.   E and K
photometers provided low- and high-sensitivity ranges respectively.
Signals marked S were saturated.   From 2 March to 14 March the
visible disk was spotless.

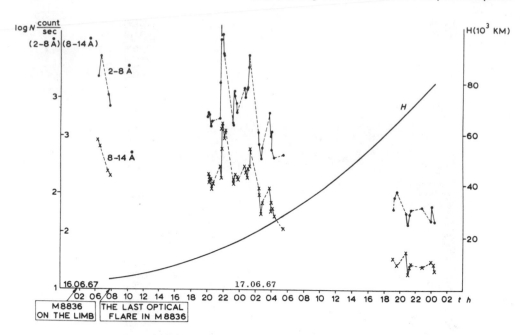

Fig 3. X-ray flux from the western limb on 16-18 June 1967 during the disappearance of the active region M 8836.

Mandelshtam[5], in another review, provided the following maximum-to-minimum tabulation:

Table 2   Solar maximum and minimum x-ray fluxes

Range	Flux (minimum)	Flux (maximum)
0-10 A	10^{-5}	2-3 x 10^{-3}
0-8 A	5 x 10^{-6}	1-1.5 x 10^{-3}
10-20 A	10^{-4}	1-2 x 10^{-2}
44-60 A	10^{-2}	5 x 10^{-2}

The behavior of a particular eclipse-ionosphere may be expected to depend on the phase of the solar cycle if the ratio of x-ray to ultraviolet radiation is important.

DISTRIBUTION OF XUV RADIATION OVER THE SOLAR DISK

The transition region from chromosphere to corona is very
abrupt, as shown in Fig. 4. Specific stages of ionization of any
given atom appear within relatively narrow ranges of temperature
so that we may expect a strongly differentiated height distribution of stages of ionization. Emission from O II to O VI and of
Si II to V should be confined to a narrow range near 2000 km above
the photosphere. Higher stages of ionization in Si, Mg and Fe,
however, are found at heights of hundreds of thousands of kilometers. From NRL rocket spectroheliograms and the raster images
produced by the Harvard experiment in OSO 4, a great deal has been
learned about the height distribution of emission of ionizing
radiation as well as limb brightening and enhancement over active
centers.

In the XUV spectrum, a series of rocket spectroheliograms
photographed by Purcell and Tousey[5] have shown clearly the spatial
distribution of most of the important ultraviolet emission lines.
The NRL technique employed a wide entrance slit or a slitless
objective grating spectrograph. For lines that are widely
separated in wavelength, the dispersion is sufficient to show well
separated spectroheliograms. Where spectral lines are closely

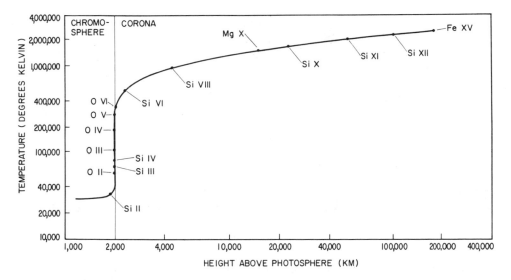

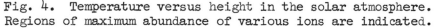

Fig. 4. Temperature versus height in the solar atmosphere.
Regions of maximum abundance of various ions are indicated.

clustered, overlapping images introduce confusion.   From the
earliest use of these techniques the major conclusions drawn were
that the solar disk exhibits little or no limb brightening in
H Lyman α and β and C III (977 A).   In contrast, O VI (1031.9 A,
1037.6 A) and Mg IX (368 A) show sharp limb brightening. With
increasing ionizing energy, lines such as Fe XV (284 A) and
Fe XVI (335 A) show emission from highly concentrated knots over
active plages. More recent observations (1966-1969) have yielded
spectroheliograms of improved resolution. Images in Figs. 5, 6
cover the wavelength range 150 to 700 A and have a spatial resolution of about 10 arc seconds. Included in Fig. 5 are Fraunhofer
Institute maps for the period 26-30 April 1966 and the arrows at
the right edge of each strip of spectroheliograms indicate the
plage positions.   He II (304 A) shows emission from the entire
disk but lines from 171 to 400 A give bright coronal rings and
localized bright spots over plages. With increasing stages of
ionization the spot emissions become more concentrated and relatively brighter. O IV (555 A) shows a limb ring almost uniformly
bright all around the circumference but Mg IX (368 A) and Ne VIII
(465 A) show gaps in the polar regions.

   A simple broad band heliograph for the range 171-400 A has
also been flown. Figure 7 shows the images obtained on 27 July
1966 using an off axis paraboloidal mirror and an aluminum filter.

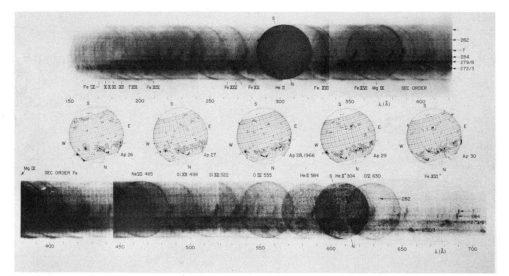

Fig. 5. Objective-grating images of the sun 150-700 A. Photographed by the Naval Research Laboratory 28 April 1966.
Fraunhofer Institute maps for the period 26-30 April 1966 are
shown for comparison.

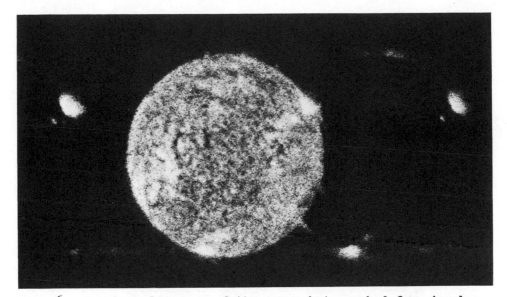

Fig. 6. Spectroheliograms of the sun, photographed from Aerobee
rocket, 9 May 1967. This small section of the spectrum shows the
solar disk in singly ionized helium (304 A), on the left, overlapping 15 times ionized iron (284 A), on the right. The helium
image shows emission from the entire disk, enhanced emission from
plages over sunspots, and a coarse network of bright and dark
mottling. The iron image shows virtually no disk radiation. All
of the emission comes from a few isolated active condensations.
These active regions account for most of the ionizing radiation
that affects the lower ionosphere. (NRL photograph)

With weak exposure the active regions dominate the image. As the
exposure is increased, general emission from high in the corona
is recorded.

The Harvard spectroheliograph on OSO-4 was designed to
record the spectrum from 300-1400 A and to produce raster images
in selected wavelengths. Figures 8, 9 illustrate the extent of
limb brightening observed from raster scans in different wavelengths. From such limb brightening curves and the measured
fluxes it is possible to calculate the eclipse variation of ionizing radiations throughout the UV spectrum.

ECLIPSE OBSERVATIONS

The development of two-stage solid fuel rocket systems, such
as the Nike Asp, made it possible at the time of the 12 October
1958 eclipse to launch a series of rockets in rapid succession

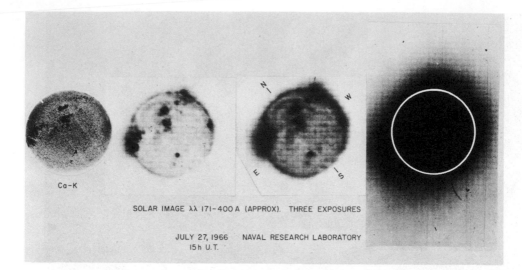

Fig. 7. Solar images in the band 171-400 A. Photographed by the Naval Research Laboratory on 27 July 1966, at three different exposures. A Ca-K spectroheliogram photographed on the same day at the McMath-Hulburt Observatory is shown for comparison at the left.

and to follow the occultation of ionizing radiation. The NRL eclipse expedition to the South Pacific succeeded in launching five rockets at times calculated to reveal the contributions of x-ray emission from discrete active plage regions on the disk and the residual x-ray flux from the uneclipsed corona at totality.

The information derived from the observations[8] was summarized as follows: (1) in the x-ray band from 44-60 A, 10 to 13 percent of the solar flux remained uneclipsed at totality; by contrast H Lyman α (1216 A) was reduced to 0.05 percent; (2) bright plage regions on the east limb yielded six times the flux observed from a comparable crescent of the disk on the west limb that did not contain any bright plages.

In summarizing the available information at the time of the Symposium on Solar Eclipses and the Ionosphere which was held in London in 1955, Ratcliffe[9] concluded that an appreciable portion of the E-region ionizing radiation, about 10 to 15 percent, came from beyond the visible disk and was not obliterated at totality. The 1958 rocket observations appeared to confirm these conclusions and indicated the importance of x-rays in controlling E region. Assuming x-rays to be the primary ionization source, the behavior

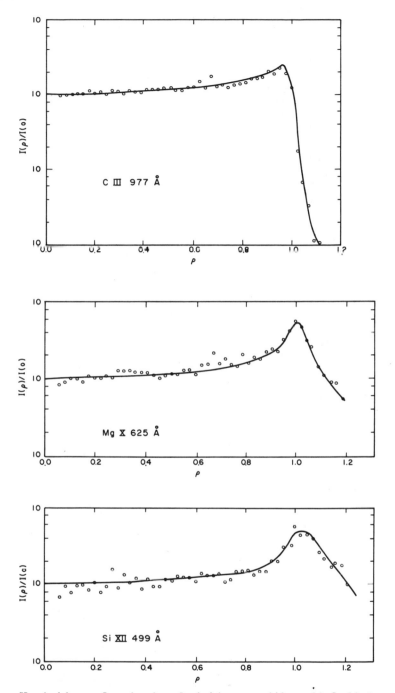

Fig. 8. Variation of emission brightness with radial distance
across the solar disk.

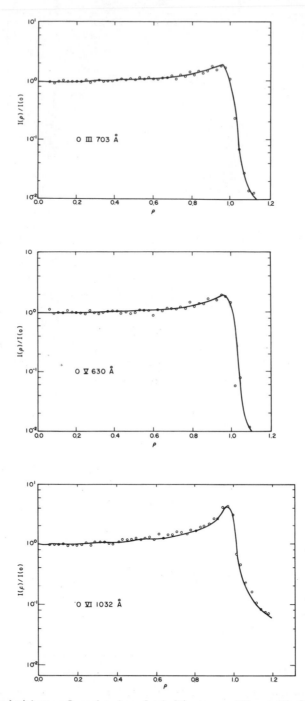

Fig. 9. Variation of emission brightness with radial distance across the solar disk.

of the residual ionosphere at totality in 1958 implied a recombination coefficient of 4×10^{-8} cm^3/sec.

The most pertinent theoretical calculations of solar x-ray emission available in 1958 were those of Elwert.[10] Assuming that the emission came from a small uniform corona distorted only slightly by a small equatorial bulge and polar flatness, he calculated the residual fluxes for various eclipses. As a function of the ratio, η, of the moon's apparent diameter to the sun's, he obtained the following percentages of residual x-ray flux at totality.

Table 3   Residual x-ray fluxes at total eclipses

Date	η	% Residual Flux
30 June 1954	1.02	17 - 22
20 June 1955	1.06	10 - 12
12 Oct. 1958	1.04	14 - 17

To derive these results Elwert computed the expected limb brightening. The corona is optically thin to continuum radiation but the optical depth, τ, becomes important for line radiation. Elwert's calculations show that limb brightening could reach a factor of about 10 in continuum radiation, a factor of about 6 for $\tau = 0.5$, and a factor of about 4 for $\tau = 1.0$ in a spherically symmetrical corona. The residual flux at totality, however, is much less dependent on the optical depth. For the case of $\eta = 1.04$, the residual flux would be 20% if the optical depth were zero, 17% if τ were 0.5, and 14% if τ were unity.

Another series of rocket launchings was successfully carried out[11] during the total eclipse of 1961 in the U.S.S.R. The Russian rockets measured the flux for wavelengths below 10 A and observed a residual of 8×10^{-5} erg/cm^2/sec at totality. Since the flux from the full disk was about 10^{-3} erg/cm^2/sec, the uneclipsed residual appeared to be of the order of 10%. It was furthermore estimated that the x-ray emission extended to heights of 50,000-100,000 km above the photosphere.

For the eclipse of 20 May 1966, the SOLRAD satellite provided data at the Arcetri Observatory in Florence, Italy and at stations at Brazzaville, Ouadagoudou, Hammaguir and Bretigny of the French National Center for Space Studies (CNES). As the moon eclipsed active centers on the disk, the x-ray emission exhibited abrupt decreases. According to Landini, Russo and Taliaferri[12] of the Arcetri Observatory, soft x-rays (8-20 A) were concentrated in coronal knots measuring less than 50 arc seconds or about 3% of the disk diameter. Harder x-rays (1-8 A) were concentrated in still smaller knots, less than 30 arc seconds in diameter.

The results of the CNES observations have been reported by G. Simon.[13] Figure 10 shows the distribution of plages. Hatched areas give the positions of the radio microwave sources and the black dots show the sunspots. Positions of the occulting edge of the moon are indicated by arcs at the start and finish of x-ray source occultations. The green corona is shown in solid line outline and the red corona by the dashed line. Figures 11, 12 show the observations of x-ray and ultraviolet fluxes during the eclipse. The northwest source was occulted at 08 h 18 m UT; the northeast at 10 h 04 m 40 s UT; and the south source at 10 h 09 m UT. Also shown is the reappearance of the northeast source between 10 h 11 m 20 s and 10 h 13 m 10 s UT. The arrows mark the beginning of the occultation of the disk. Table 4 lists the fluxes attributed to each source and Table 5 gives the full widths in seconds of arc of both the x-ray and radio sources. Source 8310 appeared to be made of a soft component extending

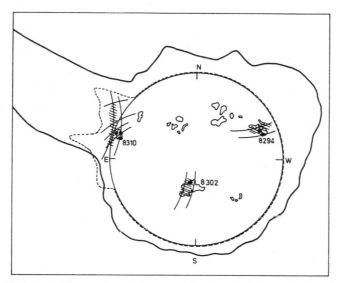

Fig. 10 Distribution of active regions on the solar disk, 20 May 1966. Calcium plages and sunspots (●) are indicated. Hatched areas show the positions of radio sources, and arcs of circles show the positions of the edge of the moon at the beginning and end of occultations or reappearances of x-ray sources. Intensities of the green and red coronal lines are indicated by solid and dashed outlines.

Table 4 Distribution of the solar flux on 20 May 1966 (10^{-4} erg/cm^2/sec)

λ Source	1-8 A Geiger Counter Flux	P	1-8 A Photometer H Flux	P	1-8 A Photometer G Flux	P	8-16 A Flux	P	0-20 A Flux	P	9.1 cm (1) P	3.1 cm (1) P	32 cm (2) P
Sun	3.60		3.65		2.76		97.0		86.0				
8310[a]			1.00	27.4	0.76	27.6	26.0	26.8	29.0	33.7	7.8	1.91	
8310[b]					0.27	9.8	10.0	10.3					
8310[c]					0.52	18.9	20.0	20.6					
8302			0.65	17.8	0.61	22.1	20.0	20.6	11.0	12.8	5.1	2.5	
8294	1.60	44.5	2.15	59.0	1.38	50.0	47.0	48.5	33.0	38.4	16.7	2.8	3.4
Σ sources			3.80	104.2	2.75	99.7	93.0	95.9	73.0	84.9	19.6	7.2	

a   Occultation
b   Reappearance of weak component
c   Reappearance of bright component
P= Percentage of total solar flux

(1) After Drago and Noci (1968)
(2) After Poumeyrol (1967)

Table 5 Angular size of solar x-ray sources and radio sources (in seconds of arc)

	1-8 Å	8-16 Å	Average Resolution	3.1 cm (1)	9.1 cm (1)	32 cm (2)	260 MHz (3)	239 MHz (4)
8310[a]	65	70	13.5	97	157		60-80	60
8310[b]	50	65	} 12.3					
8310[c]	310	341						
8302	62	64	15.2	144	160	190		
8294	75	81	12.6	136	159			

[a] Occultation
[b] Reappearance (bright component)
[c] Reappearance (total length)
(1) After Drago and Noci (1968). The resolution is approximately 10 seconds; (2) after Pourmeyrol (1967)
(3) after Letfus et al. (1967); (4) after Abrami et al. (1967)

along the limb and projecting high above the disk, and a bright
compact source directly over the plage. The emission of the bright
knot was centered about 35,000 km above the photosphere, whereas
the weaker component was centered at about 62,000 km. The total
extensions were 220,000 km (1-8 A) and 250,000 km (8-16 A).

Simon estimated that essentially all but 4% (8-16 A) and 15%
(0-20 A) of the solar x-ray emission was concentrated in the active
centers. The x-ray emitting regions appeared to be about 2.5
times smaller in diameter than the radio sources.

Working with E-layer critical frequency observations from
Sofia amd Mitchurin for the same eclipse of 20 May 1966, Taubenheim
and Serafimov[14] attempted to compute the brightness distribution
of soft x-rays. The model which they derived distributes the
effective E-layer ionization sources as follows:

Uniform disk	70%
Green-line corona	20%
Bright northwest spot	4%
Unobscured corona	3%
Radio source in northwest sector	2%
Radio source near central meridian	1%
Radio source in northeast section	0%

According to these results, the contributions of localized
sources to E region are very small. Such behavior is to be
expected if most of the E region ionization is produced by x-rays
at wavelengths longer than 20 A. Slit scan measurements of the
solar disk have, in fact, shown that at 44-60 A the entire disk is
bright and only small enhancements come from the active plages.
In contrast, radiation of wavelengths shorter than 20 A are concentrated almost entirely in the active regions that contribute
primarily to the ionization of the upper D region.

From the behavior of the eclipse ionosphere, Taubenheim and
Serafimov conclude that soft x-rays are the main source of control
of E region and that the effective recombination coefficient is
$> 8 \times 10^{-8}$ cm^3/sec.

The few examples given above are illustrative of the kinds of
observations of solar flux that can be coordinated with eclipse
observations. The next opportunity for further coordinated
studies will take place on 7 March 1970. The path of totality
will cross the Wallops Island Range of the National Aeronautics and
Space Administration in the U. S. A. and several rocket launchings
will be made during the event. Related observations will also be
made from the White Sands Missile Range in New Mexico, U. S. A. At
the same time it is expected that considerable information about
solar UV and x-ray emission will be available from presently
operating satellites.

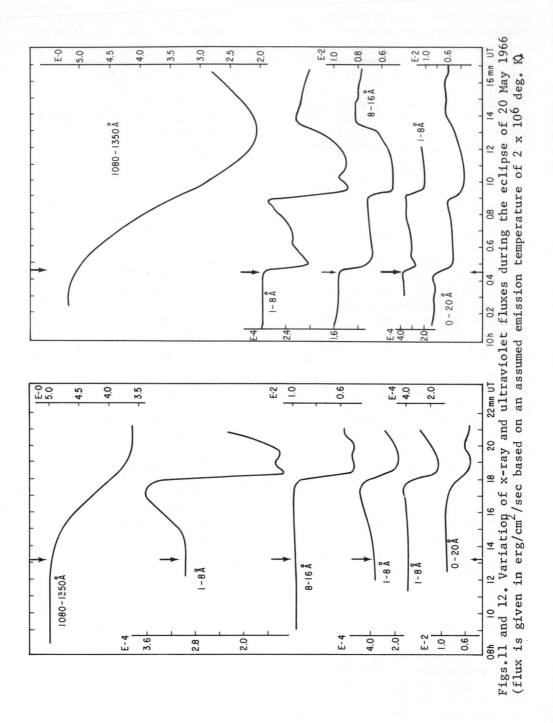

Figs. 11 and 12. Variation of x-ray and ultraviolet fluxes during the eclipse of 20 May 1966 (flux is given in erg/cm²/sec based on an assumed emission temperature of 2 x 10⁶ deg. K)

REFERENCES

[1] H. E. Hinteregger, "Effects of Solar XUV Radiation on the Earth's Atmosphere," Annals of the IQSY, Vol. 5 (The M. I. T. Press, 1969), pp. 305-321.

[2] R. L. Blake, T. A. Chubb, H. Friedman, and A. E. Unzicker, "Spectral and Photometric Measurements of Solar X-ray Emission below 60 A," Astrophys. J. 142, 1 (1965).

[3] I. L. Beigman, Yu. I. Grineva, S. L. Mandelshtam, L. A. Vainstein, and I. Zitnik, "On the Localization, Size and Structure of the Regions of the X-ray Flares on the Sun," Solar Physics 9, 160 (1969).

[4] H. Friedman, "Energetic Solar Radiations," Annals of the IQSY, Vol. 4 (The M. I. T. Press, 1969), pp. 36-56.

[5] S. L. Mandelshtam, "X-ray Emission of the Sun," Space Sci Rev. 4, 587-665 (1965).

[6] J. D. Purcell and R. Tousey, "XUV Heliograms," A. A. S. Special Meeting on Solar Astronomy, Boulder, Colorado, 3-5 October (1966).

[7] G. L. Withbroe, "Solar XUV Limb Brightening Observations and Interpretation," Harvard College Observatory Tech. Report No. 13, November 1969.

[8] T. A. Chubb, H. Friedman, R. W. Kreplin, R. L. Blake, and A. E. Unzicker, "I. X-ray and Ultraviolet Measurements during the Eclipse of October 12, 1958; II. X-ray Solar Disk Photograph," Memoires Soc. R. Sc. Liege, 5th Series, Vol. IV, pp. 228-240 (1961).

[9] J. A. Ratcliffe, "Solar Eclipses and the Ionosphere," ed. by Beynon and Brown (Pergamon Press, 1956), p. 158.

[10] G. Elwert, "Distribution of X-rays Emitted by the Solar Corona and the Residual Intensity during Solar Eclipses," J. Atmos Terr. Phys. 12, 187 (1958).

[11] S. L. Mandelshtam, Yu. K. Voronko, I. P. Tindo, A. I. Shurygin, and B. N. Vasilyev, "Investigation of Solar Emission during the Total Solar Eclipse of 15 February 1961," Dokl. Akad. Nauk SSSR, 142, 77-80 (1962).

[12] M. Landini, D. Russo, and G. L. Tagliaferri, "Solar Eclipse of May 20, 1966, Observed by the Solrad-8 Satellite in X-ray and Ultraviolet Bands," Nature 211, 393-394 (1966).

[13] G. Simon, "Le Rayonnement X du Soleil Lors de L'Eclipse du 20 May 1966," Solar Physics 7, 295-301 (1969).

[14] J. Taubenheim and K. Serafimov, "Brightness Distribution of Soft X-rays on the Sun, inferred from Ionospheric E-layer Variations during an Eclipse," J. Atmos. Terr. Phys. 31, 307-312 (1969).

X-RAY EMISSIONS FROM THE SUN

S. M. Krimigis

The Johns Hopkins University Applied Physics Laboratory

C. D. Wende, National Space Science Data Center

I.   INTRODUCTION

Since the existence of solar x-rays was inferred from the work of Grotrian [1939] and Edlen [1942], there have been in recent years several reviews of the experimental and theoretical work on the subject, including those of Mandelstam [1952, 1965], Friedman [1963], de Jager [1964], and most recently Underwood [1968]. Many of these reviews contain excellent discussions of both the experimental observations and the physical processes responsible for solar x-ray production, and some include descriptions of experimental techniques. It is only recently, however, that observational knowledge of solar x-rays has vastly increased with the massive set of observations obtained with the OSO satellite series, and the long term measurements obtained by Explorers 33 and 35, and Mariner 5 spacecraft.

In the present discussion, it is proposed to give a brief review of the observations prior to 1965, and concentrate on the most recent results, in an attempt to give as quantitative a picture of solar x-ray production, as current observational knowledge will permit. Our attention will be directed to those emissions generally originating in the solar corona ($\lambda \leqslant 100$ Å), which represent an arbitrary upper limit in wavelength, albeit a reasonable one from an experimental standpoint; in fact, measurements below 100 Å require techniques which are different than these employed above 100 Å [Boyd, 1965]. Further, the x-ray component is compared to the radio flux, in an attempt to gain some information on the production mechanisms on the sun. In part II, observations representative of the quiet sun are presented and data are shown illustrating the active sun at various time scales.

In part III a brief discussion of the relevant theory is given and
the implications of current observations in the understanding of
physical processes occurring on the sun are discussed.

Taking note of the subject of this conference, Figure 1 illustrates the importance of solar x-rays in the 1-100 Å region to the
physics of the ionosphere, particularly the D and E regions. It
can be seen that x-ray photons can easily penetrate down to the
lower limit of the D-layer and, if the energy flux is sufficient,
as appears to be the case from SID observations, influence deci-
sively the chemistry of the ionosphere. Thus, knowledge of the
details of solar x-ray flux below 100 Å is of critical importance
in the correct interpretation of associated eclipse observations.

II.   OBSERVATIONS

The first observation of solar x-rays was obtained by Burnight
[1949] using Schumann plates with aluminum and beryllium filters.
After exposure to sunlight and subsequent development, the films
showed black spots which were attributed to solar x-rays. Since
this first observation, a great deal of pioneering work in this
field was carried out at the Naval Research Laboratory (NRL) using
rockets and satellites. The early work carried out by means of
rockets will not be discussed here, but rather the more extensive
satellite data of the last few years will be concentrated upon.

1.   Quiet Sun

a. _Time Variation_. It has been long established that solar
activity, as manifested by sunspot numbers, the coronal green line,
geomagnetic observations, etc., varies with an 11-year cycle. It
is necessary in discussing solar x-rays, first to establish the
variation during a solar cycle, so that short-term observations
may be viewed in the proper context. Unfortunately, continuous
observations over a full solar cycle are not yet available, although some limits have been placed on the solar cycle variation
by use of rocket and early satellite measurements by Kreplin ct al
[1962]. They estimate that the flux of x-radiation between 1953
to 1959 increased by factors of ~ 600-1000, 60, and 7, in the
ranges 2-8, 8-20, and 44-60 Å, respectively.

A more reliable estimate of the solar cycle variation in the
2-12 Å region has recently been obtained by Wende [1969a] using
Anton Type 213 GM tubes from Injuns I and III, flown in 1961-62 and
1962-63 respectively, and Explorers 33 and 35, flown in 1966 and
1967 (for experimental details, see Van Allen, 1967). In averaging
the data, monthly time blocks were used, as they correspond to the
approximate period of one solar rotation and thus represent the

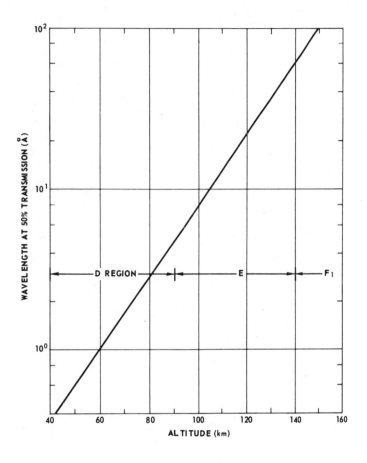

Figure 1:   X-ray penetration depth into the ionosphere.   The wavelength is given at the 50% transmission point.

level of activity of the entire sun rather than that of the visible
disk only. Figure 2 shows the observed soft x-ray fluxes (with
flare data deleted) and the Zurich sunspot number over the years
1961 to 1968. It is seen that the fluxes observed by Injun I and
Explorers 33 and 35 are about an order of magnitude higher than
those measured by Injun III. We note that the maximum variation
in efficiency between these GM tubes is less than a factor of 2,
so that the observed difference is not instrumental. Figure 3
shows the same data plotted with the 10-cm radio flux averaged in
the same manner as the x-ray data. We note that the minimum in the
radio flux generally agrees with the minimum in the x-ray fluxes
so that a solar cycle variation in the x-rays is readily apparent.

It is of interest to note that during the 1966-67 period the
quiet-sun x-ray flux was $\sim$ 2 x 10^{-3} ergs/cm^2sec. This flux is not
necessarily representative of the so-called "quiet-sun component"
which is defined by Kundu [1964] as "the emission of the solar
atmosphere when contributions from all discrete sources of the
slowly varying component are subtracted." The minimum flux observed by either Explorers 33 or 35 was 3.9 x 10^{-4} ergs/cm^2sec,
which is still greater than some of the average fluxes observed
by Injun 3, and suggests the presence of such a quiet-sun component.
An upper limit to the quiet-sun component of 6 x 10^{-4} ergs/cm^2sec
was given by Friedman [1964] for the flux below 8 $\overset{\circ}{A}$, from observations obtained during the period 13 July to 3 August 1960.

The solar cycle variation for several ranges in wavelength
were summarized by Mandelstam [1965] using observations obtained
during the 1960-64 period. These are as follows:

$$< 10 \ \overset{\circ}{A} \approx 10^{-5} \text{ to } \sim 2 \text{ x } 10^{-3} \text{ ergs/cm}^2\text{sec}$$
$$10 - 20 \ \overset{\circ}{A} \approx 10^{-4} \text{ to } \sim 10^{-3} \text{ ergs/cm}^2\text{sec}$$
$$44 - 60 \ \overset{\circ}{A} \approx 10^{-2} \text{ to } \sim 5 \text{ x } 10^{-2} \text{ ergs/cm}^2\text{sec}$$

We observe that as the wavelength increases the solar cycle variation becomes less pronounced. In addition, the < 10 $\overset{\circ}{A}$ limits agree
well with the more extensive observations of Explorers 33 and 35
and Injuns I and III.

b. Energy Spectrum. The spectrum of solar x-rays is generally
divided into (a) the continuum and (b) line structure. Observations
were initially restricted to the continuum, mainly because of lack
of adequate resolution in the detecting apparatus. A typical (and
much used) example obtained with a proportional counter flown on
Ariel-1 is shown in Figure 4 [Culhane et al, 1964]. We point out
the curve labeled "11.41 - 11.50" which represents the spectrum of
the quiet sun. In efforts to describe the spectrum in a quantitative manner, Friedman et al [1951] assumed that the energy distribution in the solar spectrum may be described as that of a grey

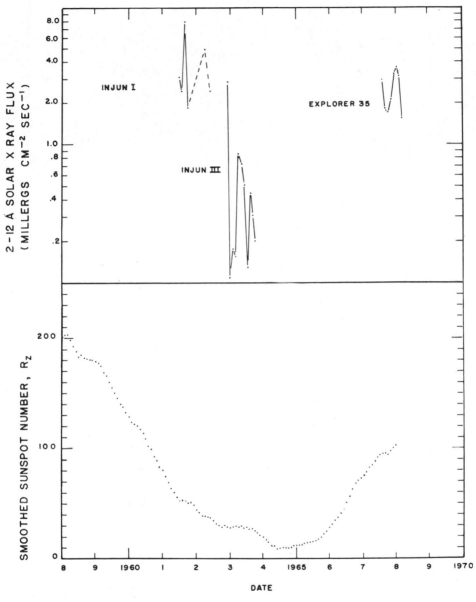

Figure 2:   The solar cycle variation observed in the x-ray and
             Zurich sunspot number. The fluxes are averaged in
             monthly intervals.

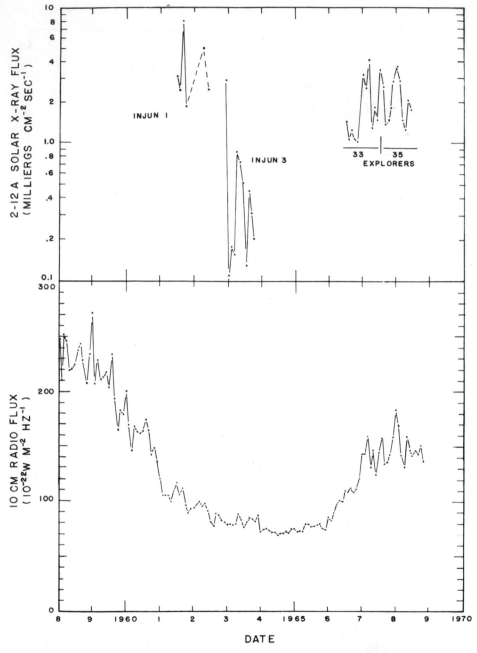

Figure 3:   The same as in Figure 2, but with the 10-cm flux re-
placing the Zurich sunspot number [Wende, 1969a].

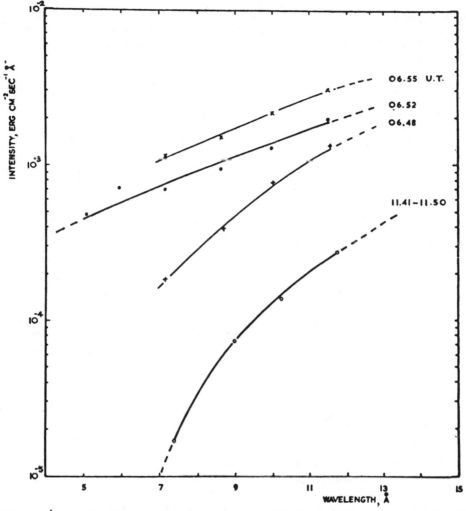

Figure 4:   Spectrum of solar x-rays obtained with a proportional
            counter on Ariel-1. The spectrum hardens during a
            class-2 flare on 3 May 1962. The quiet sun spectrum
            is labeled "11.41-11.50", [Culhane et al. 1964].

body, i.e.,

$$J(\nu) \propto \frac{1}{\nu^3} \quad \frac{1}{e^{h\nu/kT}g - 1}$$

It is generally found that T_g has different values for different
spectral ranges. For the example given above, T_g is in the range
$\sim 2 \times 10^6$ °K, similar to the temperature of the solar corona. We
shall come back to this figure when we discuss spectral changes
during flares.

Figure 5 shows the spectrum of solar x-rays in the range 7 to
~ 20 Å [Neupert et al, 1967]. The lower curve shows a quiet sun
spectrum obtained on 21 March 1967. There are several emission
lines at wavelengths longer than 14 Å, which represent high states
of ionization of iron. These lines probably originate in solar
active regions and imply temperatures of several million degrees.
The spectrum of solar x-rays shall be treated in more detail in
the discussion of solar flares.

The soft x-ray emission from the quiet sun exhibits limb
brightening [Friedman, 1964; Reidy et al, 1968; Underwood, 1968].
We show here an example in Figure 6 (courtesy of Dr. Underwood)
obtained with a rocket flight on 12 November 1966 at 1620 U.T.
The bandpass of the filter was 3-22 Å and 44 to 60 Å and a 10-
second exposure was used. Limb brightening is clearly visible at
these wavelengths, and by comparing this picture with a similar one
obtained with a beryllium filter we conclude that the radiation is
in the 44-60 Å band. This observation is not surprising in that
limb brightening is a characteristic of optically thin atmospheres.

2. Active Sun

It is generally accepted that the 11-year intensity variation
in the x-ray wavelengths is due to the relative number of active
regions appearing on the sun between solar maximum and solar minimum. In fact, even at solar minimum there are usually a few active
regions on the sun that are visible in H_α or the calcium K line.
Typically, the enhanced radio and x-ray emissions associated with
a plage area originate from regions of enchanced electron density
and temperature in the overlaying corona.

a. Source Regions. A series of x-ray photographs of the solar
disk obtained by Underwood [1968], using a grazing incidence telescope on a rocket flown on 20 May 1966, is shown on Figure 7. A
Fraunhofer map of the sun on the same day is shown for comparison.
In Figure 7a, the wavelength band is in the range 3-11 Å and one
observes that the emission is confined to three well-defined areas,
having equivalent counterparts on the Fraunhofer map. It can also

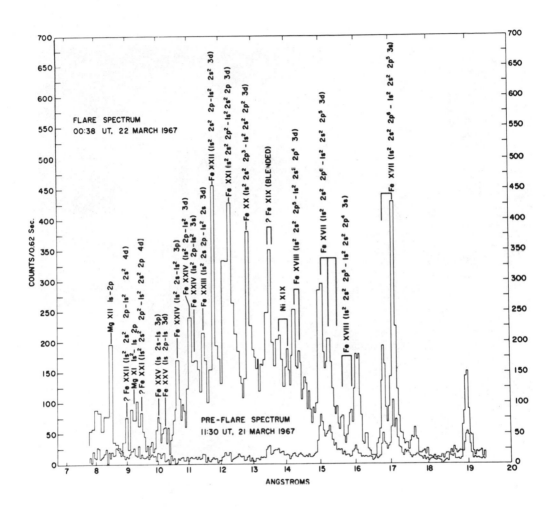

Figure 5:　Comparison of the solar spectrum between 6.3 and 20.0 Å obtained during a flare on 22 March 1967, with a spectrum obtained on the previous day when no flares were in progress [Neupert et al, 1967] (courtesy of Dr. Neupert).

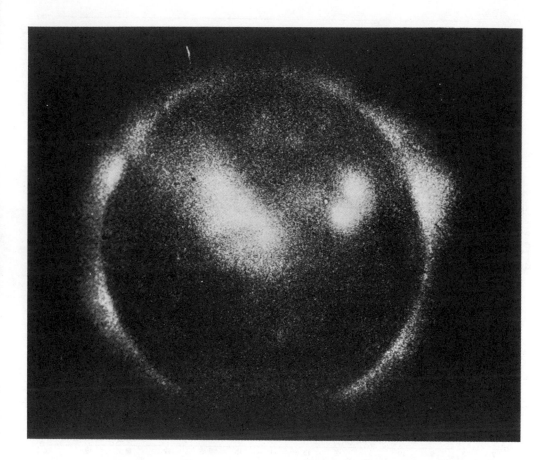

Figure 6:   X-ray photograph of the sun obtained on 12 November
             1966 at ∼ 1620 U.T. The wavelength is 44-60Å (contami-
             nated by some radiation in the 2-21 Å range) with an
             exposure of 10 seconds. Note the lack of limb brighten-
             ing over the south pole (courtesy of J.H. Underwood)
             [Underwood, 1968].

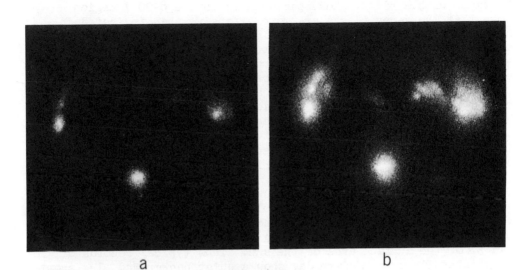

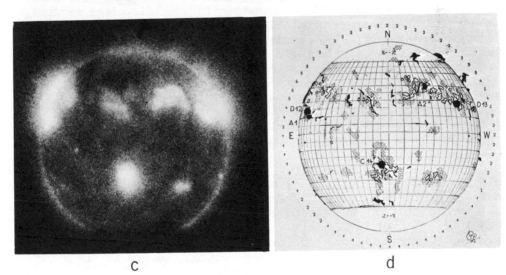

Figure 7:   X-ray photographs of the sun obtained on 20 May 1966.
(a) Wavelength region 3-11 Å; (b) Wavelength region
8-20 Å; (c) Wavelength region 27-40 Å (contaminated
by some radiation at 3-11 Å); (d) Fraunhofer Institut
map of the sun on the same day.

be seen that there is some x-ray emission on the northeast limb,
not associated with a visible plage region indicating that the
source of x-radiation is probably high in the solar corona. Figure
5b, taken with a filter whose passband is in the 8-20 Å region,
shows that the x-ray plages are more extensive than indicated in
5a and correspond closely to the bright calcium regions. Finally,
in 5c a photograph taken with a filter having a passband in the
3-11 and 27-40 Å ranges is shown, and by comparing it to 5a, one
sees that the picture basically represents the 27-40 Å passband.
One notes that at those wavelengths all calcium plage regions have
their x-ray counterparts. In addition, limb brightening around
the disk with the exception of the south pole may be seen. This
observation is in line with the picture shown in Figure 6, and indicates that the electron density decreases as one approaches the
south pole with the transition occurring $\sim 60°$ latitude.

b. Temporal Variation. The slowly varying component of the
soft x-ray emission is a direct result of changes in intensity
over plage regions. As seen from the previous pictures, probably
80% of the emission at small (< 20 Å) wavelengths comes from less
than 5% of the surface area of the sun. Thus typical records of
x-ray intensity over a given day show a quiet background plus short-
term increases most probably originating in specific plage regions.
Two days of observations obtained with a GM tube on Mariner 5,
having a passband of 2-9 Å [Wende, 1969] are shown in Figure 8.
The top panel shows a relatively active day with several short-term
(~ 1 hour) increases corresponding to small flares on the sun; we
define the baseline of intensity as the intensity of the slowly-
varying component for that day ($\sim 1.5 \times 10^{-3}$ ergs/cm^{2}sec on 30 July
1967). The bottom panel shows the level of x-ray activity on
24 June 1967; note that the baseline of intensity for this day is
approximately an order of magnitude less than on 30 July, at
$\sim 3 \times 10^{-4}$ ergs/cm^{2}sec.

To investigate further how the slowly varying component behaves,
averages of the non-flare flux from Mariner 5 and the corresponding
10-cm radio flux for the period 14 June to 21 November 1967 [Wende,
1969a] are plotted in Figure 9. The x-ray flux varies by at least
an order of magnitude over the period of observations, and, on the
whole, there is a general association between changes in the x-ray
and 10-cm radio flux.

Periods of enchanced activity occur during solar rotations
1833, 1834, and 1836, and each enchancement can be associated with
a specific active region on the sun. The enchancement of activity
beginning around 20 July was due to a series of active regions
which were rotated onto the northeast limb beginning on 21 July
[Solar Geophysical Data, 1967]. The most important active region
was a large and complex spot group located at N27 W02 on 28 July
(McMath plage number 8905) which completed its west limb passage

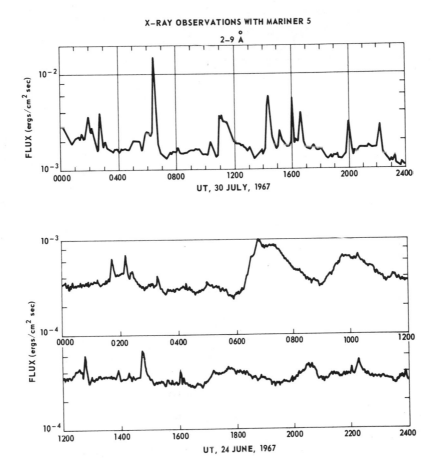

Figure 8:   X-ray observations from Mariner V on 30 July and 24
June 1967.   The level of activity for the two days
differs by almost an order of magnitude.

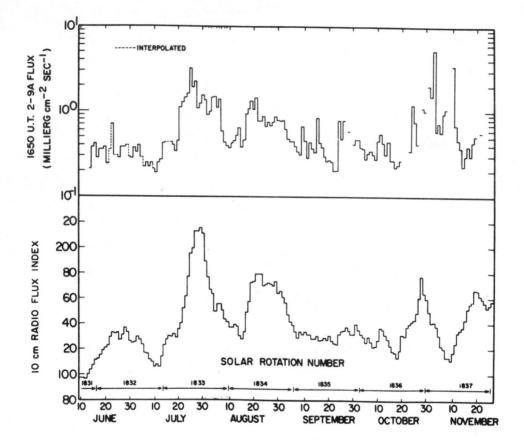

Figure 9:   The history of Mariner V x-ray fluxes and 10-cm radio
            fluxes observed in 1967 [Wende, 1969a].

on 4 August. The drop-off in the x-ray intensity about three days
later is quite abrupt. This information will be used later to
estimate the height of the x-ray emitting region as being > 50,000
km. Similarly, the activity centering around 25 August can be
associated with plage region number 8942 at N22, which is believed
to be the reappearance of plage number 8905.

Note that although the 10-cm radio flux clearly tracks the
slowly varying x-ray component most of the time, there is a discrepancy between the two beginning in late October. This is due to
the fact that the difference in heliocentric longitude between the
earth and Mariner 5 was about 40 degrees at that time, so that
Mariner 5 could observe a given feature on the solar disk for at
least three days after it had disappeared behind the west limb as
viewed from earth. To investigate further this effect, Figure 10
shows the daily x-ray flux as observed by Explorer 35 in the vicinity of earth (2-12 $\mathring{A}$) in October and November 1967. Physically
one expects that the 2-12 $\mathring{A}$ flux (Explorer 35) should always exceed
the 2-9 $\mathring{A}$ flux (Mariner 5). It is noted however that in late
October and early November this condition is not satisfied, suggesting that the Mariner detector is seeing events not visible to
Explorer 35. The observations of Mariner 5 assure that the decrease
in the intensity as seen from earth is indeed due to the disappearance of the plage behind the limb rather than the normal decay of
the region. From the time that it takes for a source to disappear
from view, the height of the emitting region can be estimated,
assuming that the photosphere is optically thick at the observing
wavelength while the chromosphere and corona are optically thin.
It can be seen that it took approximately 2 days for the flux
associated with plage 9034 (late October) to decrease to one half
its prelimb passage value, so that the apparent height of the
active region is $\sim 7.6 \times 10^4$ km. A similar analysis for plage
8905 which disappeared behind the west limb around 1 September
gives 6.8×10^4 km. Thus, simultaneous observations from two
different spacecraft separated by a large angle in heliocentric
longitude have allowed an estimate of the height of the x-ray
emitting region to be made.

c. Spectrum. As the slowly varying component is generated
by abnormally hot regions in the solar atmosphere, the x-ray
plage, it would be expected that the spectrum would harden during
periods of higher solar activity. Such a hardening of the spectrum
is illustrated on Figure 11 [Wende, 1969]. The data are one-hour
averages of the observed flux, taken every four hours, from 27 July
1967 until 10 September 1967. Points which included x-ray flares
with a peak to background flux ratio of three or greater are indicated by an "x". Note that if the non-flare data on the right-
hand half of the figure are correlated, the regression line would
not intercept the origin, but that if the data on the left-hand
portion are included the regression line curves somewhat and does

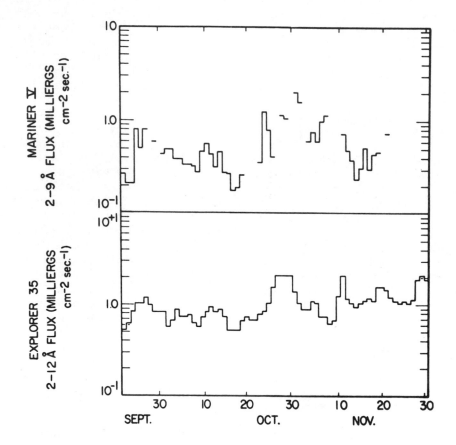

Figure 10:   Mariner V and Explorer 35 x-ray fluxes observed in the
             fall of 1967 [Wende, 1969a].

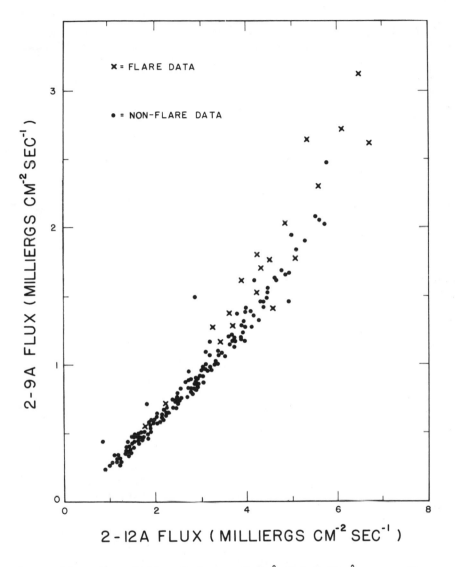

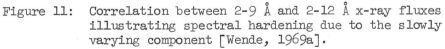

Figure 11: Correlation between 2-9 Å and 2-12 Å x-ray fluxes illustrating spectral hardening due to the slowly varying component [Wende, 1969a].

intercept the origin. Thus, the spectrum is slightly harder when
x-ray plages are present.

3. X-Ray Flares

In addition to the slowly varying component of solar x-ray
flux, rapid changes in the intensity are often observed with time
scales of a few minutes to a few hours. Most of these events are
typically associated with chromospheric flares visible in H_α, which
also produce energetic particles reaching the earth.

a. Source Regions. An importance 1N flare on 8 June 1968 was
photographed by Vaiana et al [1968]. Figure 12 shows a picture of
the flare, which was taken at the time of the H_α maximum, at
several wavelengths. The flaring region was located in the south-
east quadrant, close to central meridian. Figure 12a represents a
6-second exposure in the 3.5 to 14 Å and 44 to 60 Å region and is
somewhat overexposed. Figure 12d shows a 2-second exposure of the
same region in the 3.5 to 14 Å range. One observes that in this
wavelength range, the flare region is at least a factor of 10 more
intense than any other region on the surface of the sun, and consists of two main structures, each several minutes of arc in length.
By comparing the structure to the H_α image (12b), one observes a
striking similarity in the two pictures. Vaiana et al [1968] state
that the correspondence between the H_α and x-ray pictures is quite
extensive down to the smallest detail.

It may be seen from 12a, b, and c that all parts of the solar
disk visible in H_α are also seen in the x-ray pictures, although
the converse is not necessarily true. For example, there is a
loop structure on the southeast extending over 150,000 km above
the limb which is clearly visible in x-rays but has no counter-
part in H_α. The loop must clearly originate from a plage region
behind the limb, which in fact rotated into the visible disk about
2 days later. Another notable feature is seen towards the limb
in the southwest, where looping and interconnecting structure can
be distinguished in x-rays, where no similar structure is observable in H_α. The x-ray loops extend more than 100,000 km above the
H_α plage.

The most puzzling feature of the observations described above
is the detailed correspondence between the x-ray and H_α images of
the flare, since it is known that there is a large difference in
the degree of ionization required to account for the x-ray and the
H_α emissions. Vaiana et al [1968] suggest a strong magnetic field
connection between the two regions to account for the similarity in
shape. In regard to the active regions, it is seen that the x-ray
emitting regions resemble the H_α at the base of the corona but
that at higher levels, the x-ray regions exhibit loop-like

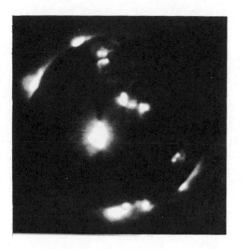

MICRON ALUMINIZED MYLAR FILTER

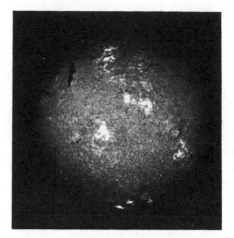

SIMULTANEOUS Hα PHOTOGRAPH
(COURTESY OF ESSA)

E ⭠ ⭡ N

**SLITLESS SPECTROGRAM
12 MICRON BERYLLIUM FILTER**

12 MICRON BERYLLIUM FILTER

Figure 12:   X-ray picture of an 1N flare taken on 8 June 1968.
(a) Wavelength region 3.5-14 Å and 44 to 60 Å with a
6-second exposure; (b) H_α image of the sun, taken on
the same day; (c) Slitless spectrograph picture in the
3.5-14 Å range; (d) 2-second exposure in the 3.5-14 Å
region [Vaiana et al, 1968] (courtesy of Drs. Vaiana
and Zehnpfennig).

structures interconnecting H_α regions, apparently determined by the magnetic field.

b. <u>Temporal Profile</u>. In discussing the x-ray flare phenomenon, the distinction must be made between the quasi-thermal emissions occurring at $\lambda > 1\text{Å}$ and the non-thermal emissions occurring at $\lambda < 1\text{Å}$ [de Jager, 1965].

A typical thermal x-ray flux is illustrated on Figure 13 [Wende, 1969b], together with the "gradual rise and fall" microwave burst which accompanied it. This event was also accompanied by three burst impulsive microwave flares, illustrated on the insert.

Except for the data obtained during the impulsive microwave events, the x-ray and microwave fluxes were highly but non-linearly correlated.

The observed correlation is consistent with a thermal flare model in which the flaring region is optically thick at the radio wavelengths (i.e., the radio flux is proportional to the temperature) and the x-rays emanate with an emissivity, dE/d_ν, proportional to exp $(- h_\nu/kT)$. The dominant variable is the change in temperature, δT. If the fractional change in temperature, $\delta T/T_p$, is known, the peak temperature T_p, and the flare solid angle (or effective diameter) can be determined from the flux correlation [Wende, 1969b]. If the temperature doubled during the course of the flare, typical peak temperatures were less than 5×10^6 degrees Kelvin and typical effective diameters were 20 - 40 arc seconds. Both emissions were assumed to originate in a common volume of gas which had a uniform temperature throughout.

Figure 14 illustrates the thermal x-ray emissions observed during the 3B flare of 8 July 1968 [Van Allen and Wende, 1969]. The 1.95 cm radio emission is also shown. The peak integral x-ray flux, occurring at 1723 GMT, is ~ 0.4 ergs/cm^{2}sec for the 2-12 Å region, making this one of the largest soft x-ray events ever observed. The time integral of the flux in excess of the preflare background was 920 ergs/cm^2 between 1707 and 2109. The onset of the x-ray event was simultaneous ($\pm$ 1 minute) with that observed at 1.95 cm. The soft x-ray flux showed a relatively smooth rise and fall, whereas the 1.95 cm flux exhibited four pulsations at roughly ten minute intervals. The pulsations may be due to non-thermal processes whose fluxes are superimposed upon a thermal component which mirrored the soft x-ray event.

At shorter wavelengths the choice between using thermal and non-thermal mechanisms becomes less clear. Kundu [1964] showed that hard x-ray fluxes during flares correlated quite well with the impulsive microwave fluxes which were attributed to non-thermal

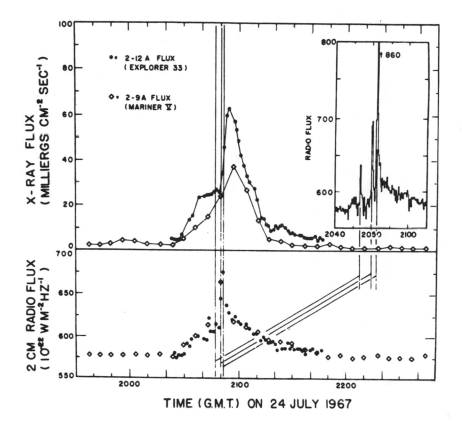

Figure 13: A 1B flare observed on 24 July 1967. On the main panels the 2-cm data points correspond temporarily to the
x-ray data points. The impulsive 2-cm bursts which
accompanied the gradual rise and fall x-ray and microwave events are illustrated with a time resolution of
10 seconds on the inset. The three vertical lines on
the main panels indicate exactly the times of the
three impulsive bursts, [Wende, 1969b].

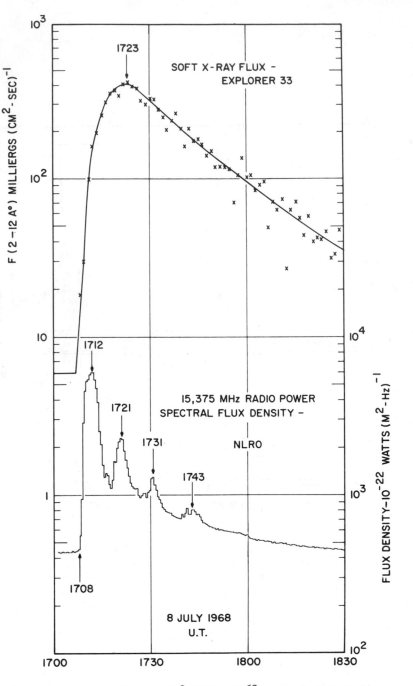

Figure 14: The large flare of 8 July 1968. Peak intensity
            reached 0.4 ergs/cm² sec [Van Allen and Wende, 1969].

processes such as synchrotron emission. Chubb et al [1966] and
Hudson et al [1969] have found, however, that at photon energies
less than about 100 kev the observed hard x-ray spectra are consistent with a flare model in which a plasma of a temperature of
the order of 10^7 to 10^8 degrees Kelvin emits x-rays as it cools
down. In the photon energy range of 80 to 500 kev, Cline et al
[1968] have shown an example of a high correlation between the
x-ray and the radio fluxes. Their data is illustrated on Figure
15. Note that intensity maxima occur at 0027, 0029, and 0037 U.T.
at both wavelengths. The theoretical explanation for the flux
correlation was given by Hold and Ramaty [1968] and will be discussed in a later section.

c. Spectrum. The spectral changes which occur in the x-ray
emissions of a flare are best dealt with in two regimes, a low
energy one (soft x-rays, with wavelengths longer than about 1 Å
and energies less than 12 kev) and a high energy one (hard x-rays,
with wavelengths shorter than about 1 Å and energies greater than
12 kev).

The gross changes in the spectrum of the soft x-ray flares
were first observed by Culhane et al [1964] by using proportional
counters. A typical flare is illustrated on Figure 4. During the
course of the flare, the spectrum clearly hardened with respect to
the pre-flare spectrum. This hardening is consistent with the
thermal or quasi-thermal model in which the temperature increased
during the course of the flare, and then decreased.

Changes in the emission line structure for wavelengths longer
than 1 Å have been observed through the use of satellite-borne
Bragg spectrometers [Meekins et al, 1968; Walker et al, 1969;
Neupert et al, 1967]. Returning to Neupert's data (Figure 5), it
is observed that not only did the continuum flux increase appreciably, but lines appeared at shorter wavelengths than in the
preflare spectrum. These have been tentatively identified, and
it is believed that ionization states as high Fe XXV are seen.
Such high ionization states are indicative of regions of very
high temperatures, typically of the order of 10^7 to 10^8 °K, pre-
suming that the concept of temperature is still valid during the
flaring process. Figure 16 shows the spectrum around 2 Å during
a flare [Neupert et al, 1967]. Note that emissions lines at
wavelengths shorter than 3 Å must originate from elements having
Z ⩾ 18 as the hydrogen like ions of the lighter elements have
line emissions only at longer wavelengths. Neupert et al [1967]
found that lines of Fe XXIV and Fe XXV are present during some
flares, and that emissions from the highest stages increase
rapidly during the flare onset, while the lower stages are observed
later in the event. They speculate that the initial ionization
produces Fe XXV (1.87 Å), and the "subsequent recombination of the
ions produces successively lower stages of ionization whose spectra

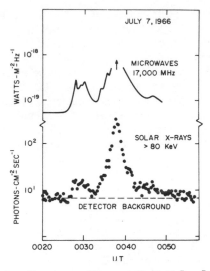

Figure 15:   The hard x-ray flare of 7 July 1966. Note the
             detailed correlation between the x-ray and radio
             fluxes [Cline et al, 1968].

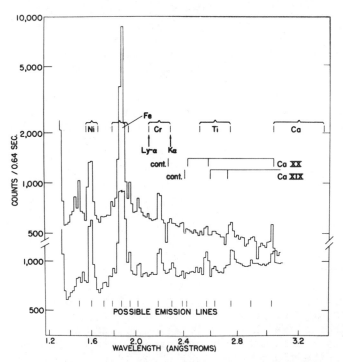

Figure 16:   Spectral scans in the region 1.3-3.1 Å obtained during
             the increasing phase of a solar x-ray burst on 22 March
             1967 [Neupert et al, 1967].

can be excited by electron impact." Although the lines in the
neighborhood of 3 Å were prominent, it is estimated that the
continuum was approximately an order of magnitude more intense
than the line emission.

In dealing with scintillation counter results in the energy
range of tens to a hundred kev, both Chubb et al [1966] and Hudson
et al [1969] have shown that their results are consistent with a
thermal spectrum providing that the temperature is higher than 10
million degrees Kelvin. It is assumed that free-free emission, or
thermal bremsstrahlung, is the dominant emission mechanism.

Cline et al [1968] showed that their spectrum for the energy
range of 80 to 500 kev x-rays from the flare of 7 July 1966 is inconsistent with any single temperature thermal flare model (i.e.,
that it is probably indicative of non-thermal mechanisms). Their
results are illustrated in Figure 17, where it can be clearly seen
that no single temperature can fit all the data.

III.   THEORY OF SOLAR X-RAY EMISSIONS

Several authors have considered the problem of solar x-ray
emission, in the framework of a high temperature in the emitting
layers, namely the corona; the calculations of Elwert [1961] are
shown in Figure 18 and are seen to be in general agreement with
the observations presented in Section II. The theoretical problem
is generally divided into thermal and non-thermal emissions.

a. Thermal.. The problem of calculating the thermal x-ray
emission from the sun is best divided into two parts. The first
part of the problem is addressed to determining the ionization
balance for each volume element in the solar atmosphere. This
balance is a function of the relative particle abundance, the
density, and the temperature of each volume element. Atoms are
ionized primarily through collisional ionization, but recombination
can occur either via radiative recombination or via dielectronic
recombination. At temperatures lower than about 100,000 degrees
Kelvin, radiative recombination predominates, but at higher temperatures dielectronic recombination predominates. For a helium
plasma with a kinetic temperature of greater than one million
degrees Kelvin, the dielectronic recombination rate coefficient
is about twenty times the radiative recombination rate coefficient
[Burgess, 1964]. Assuming particle abundances, temperatures, and
densities, and using theoretically determined cross sections, the
ionization balances, or densities of each ion in a given stage of
ionization, in principle, can be calculated.

Four processes give rise to solar x-ray emissions. These are
free-free emission, or thermal bremsstrahlung, free-bound emission,

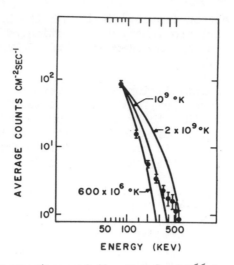

Figure 17: Energy spectrum of the 7 July 1966 hard x-ray event.
            Note that a grey body distribution does not fit the
            data [Cline et al, 1968].

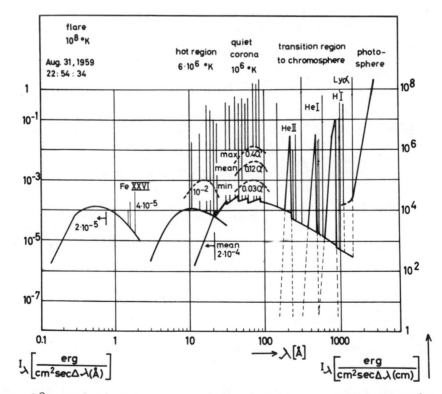

Figure 18: Calculated x-ray spectrum of the sun [Elwert, 1961].

arising from radiative recombination, bound-bound-emission, or
classical line emission, and dielectronic recombination emission.
The first two mechanisms yield continuum emission, while the latter
two give rise to line emission. The solar atmosphere is optically
thin for x-ray emissions, and thus the ion densities must merely
be inserted into the expressions for the emissivity due to each of
the above mechanisms, and the volume emissivities must be integrated over the volume of the atmosphere of the sun visible from
earth. The necessary emissivities have been given by Karzas and
Latter [1961] for free-free emissions, Elwert [1952] for free-
bound emission, Seaton [1964] for dielectronic recombination
emission (although it probably is not observable due to the weak-
ness of the emission for each individual line), and Allen [1965]
for line emission.

Elwert did the first calculation of the expected x-ray spectrum
[1961]. Mandelstam [1965] made a more comprehensive calculation,
but did not include the effects of dielectronic recombination.
This omission resulted in a free-bound to free-free emission ratio
of approximately 20. Recent calculations of Walker [private com-
munication] which included dielectronic recombination yielded a
free-bound to free-free emission ratio of about 3. A similar
calculation by Beygman and Vaynshteyn [1967] showed that the
continuum for $\lambda > 17$ Å was due primarily to bremsstrahlung by H^+
and He^{2+} ions, while for $\lambda < 17$ Å, about half the intensity is
due to recombination of the O^{7+} ion for temperatures of 1 to 4
million degrees. In all calculations, the line intensities contribute a total of less than 10 percent to the overall x-ray flux.

b. <u>Non-thermal</u>. Holt and Ramaty [1969] have examined the
flare of 7 July 1966 in great detail, with particular emphasis on
the 80-500 kev results of Cline et al [1968]. They found that by
applying the detailed theories of gyro-synchrotron emission and
absorption in a magnetoactive plasma, x-ray production by bremsstrahlung by non-thermal electrons on ambient hydrogen, and
electron relaxation in a partially ionized and magnetized gas,
they could explain the high correlation between the very hard x-ray
flux and the impulsive microwave flux in terms of source mechanisms
using a common reservoir of non-thermal non-relativistic energetic
electrons.

While Cline et al [1968] had fitted their x-ray data to an
appropriate power law in the electron flux, Holt and Ramaty [1969]
refitted the data to a power law in the electron density with no
reduction in goodness of fit (see Figure 19). This was necessary
as the radio fluxes are density dependent.

Holt and Ramaty [1969] concluded that high energy electrons
escape from the flare region, but do not easily get into the interplanetary medium. Electrons of energy $< mc^2$, however, were trapped

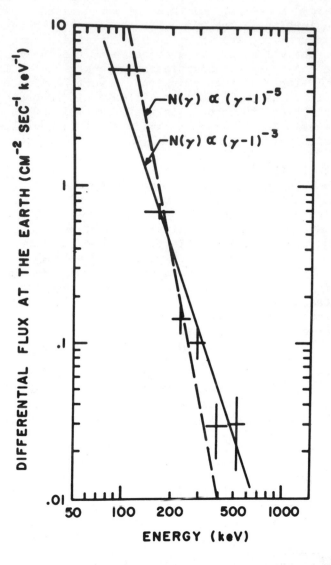

Figure 19: A fit to the spectrum of the hard x-ray event of
            7 July 1966, according to the theory of Holt and
            Ramaty [1968].

and remained to emit x-rays through collisions and microwaves and
x-rays through bremsstrahlung and gyro-synchrotron emission.

It should be mentioned that electrons of the appropriate
energy range (i.e., > 40 kev) have been observed to escape from
flares into interplanetary space [Van Allen and Krimigis, 1965].

IV.   CONCLUSION

The size of the dominant emitting regions in the sun becomes
smaller as the wavelength of observation in the x-ray region becomes shorter.   In the 44-60 Å range, the plages are relatively
diffuse and limb brightening is evident, whereas in the 1-10 Å
range the dominant sources of solar x-rays are the small ($\leqslant$ 1')
hot core regions of x-ray plages.   Thus, the source of the x-
radiation affecting the E-region of the ionosphere will be
relatively smoothly eclipsed, while the change in the radiation
affecting the D-region will be quite abrupt ($\approx$ 80% as a 1' is
eclipsed) and depend more strongly on the location and intensity
of the x-ray plage [e.g., Anastassiadis and Boviatsos, 1968].
Thus, it would be advisable to arrange for rocket or satellite
"x-ray photographs" of the sun to be made the day of the eclipse.

Secondly, should a flare occur during the eclipse, in the
λ < 20 Å region the flux from the flare could easily equal or
surpass the flux from the rest of the sun.   It would be imperative
to know the location, the temporal profile, and the spectrum of
the flare in order to ascertain its effects upon the eclipse
measurements.

Finally, one should not neglect non-x-ray sources of ionization in the D-region such as precipitating electrons [Friedman,
1964].   Recent results show that a significant flux of 40 kev
electrons continuously precipitates into the atmosphere down to
latitudes of 45° [Fritz, 1968].   Potemra and Zmuda [1969] have
computed the electron-ion pair production rate due to these
electrons and their results are shown in Figure 20.   One notes that
at night, precipitating electrons constitute the most important
source of D-region and perhaps E-region ionization.   Potemra and
Zmuda [1969] have compared their results to observed electron density profiles in the D-region and find that (Figure 21) precipitating electrons adequately explain the electron density profile
in the altitude range of ~ 50 to 90 km.   It appears then, that
during eclipse observations, satellite monitoring of precipitated
electrons is necessary for proper understanding of the D-region
ionization profile.

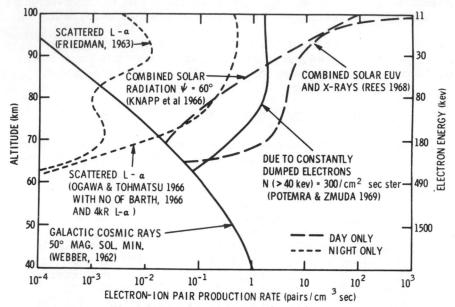

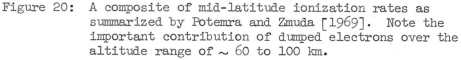

Figure 20: A composite of mid-latitude ionization rates as
            summarized by Potemra and Zmuda [1969]. Note the
            important contribution of dumped electrons over the
            altitude range of ~ 60 to 100 km.

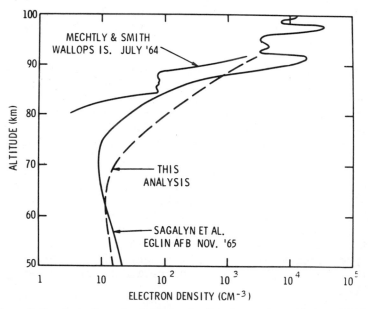

Figure 21: Calculated and observed electron density profile of
            the D-region. It appears that dumped electrons are
            the single most important contributor to D-region
            ionization down to latitudes of ~ 45° [Potemra and
            Zmuda, 1969].

REFERENCES

Allen, C. W., The interpretation of the XUV Solar Spectrum, Space Sci. Rev., 4, 91, 1965.

Anastassiadis, M. and D. S. Boviatsos, Distribution of x-ray emission from the sun deduced from measurements of ionospheric absorption, Nature, 219, 1139-1141, 1968.

Beygman, I. L. and L. A. Vaynshteyn, X-ray emission from the solar corona, Akademiya Nauk USSR, Fizicheskiy Institut Imem P. N. Lebedev Preprint No. 104, Moscow, 1967.

Boyd, R. L. F., Techniques for the measurement of extra-terrestrial soft x-radiation, Space Science Reviews, 4, 35, 1965.

Burgess, A., Dielectronic recombination and the temperature of the solar corona, Astrophys. J., 139, 776, 1964.

Burnight, T. R., Soft x-radiation in the upper atmosphere, Phys. Rev., 76, 165, 1949.

Chubb, T. A., R. W. Kreplin, and H. Friedman, Observations of hard x-ray emission from solar flares, J. Geophys. Res., 71, 3611, 1966.

Cline, T. L., S. S. Holt, and E. W. Hones, Jr., High energy x-rays from the solar flare of July 7, 1966, J. Geophys. Res., 73, 434, 1968.

Culhane, J. L., A. P. Willmore, K. A. Pounds and P. W. Sanford, Variability of the Solar x-ray spectrum below 15 Å, Space Research IV, 741, 1964.

de Jager, C., Solar ultraviolet and x-ray radiation, Research in Geophysics, I, edited by H. Odishaw, p. 1, M.I.T. Press, Cambridge, 1964.

de Jager, C., Solar x-radiation, Ann. Astrophys., 28, 125, 1965.

Edlen, B., Z. Astrophys., 22, 30, 1942.

Elwert, G., Uber die ionisations und rekombinationsprzesse in einem plasma und die ionisationsformel der sonnenkorona, Z. Naturforschg., 7a, 432, 1952.

Elwert, G., Theory of X-ray emission from the sun, J. Geophys. Res., 66, 391, 1961

REFERENCES (Cont.)

Friedman, H., S. W. Lichtman, and E. T. Byram, Photon counter
    measurements of solar x-rays and extreme ultraviolet light,
    Phys. Rev., 83, 1025, 1951.

Friedman, H., Ann. Rev. Astron. Astrophys. 1, (ed. by Leo Goldberg),
    Annual Reviews, Inc., Palo Alto, 1963.

Friedman, H., Ionospheric constitution and solar control, Research
    in Geophysics, I, 157, (ed. by H. Odishaw), M.I.T. Press,
    Cambridge, 1964.

Fritz, T. A., High latitude outer-zone boundary region for $\geq$ 40-kev
    electrons during geomagnetically quiet periods, J. Geophys.
    Res., 73, 7245, 1968.

Grotrian, W., Naturwissenschaften, 27, 214, 1939.

Holt, S. S. and R. Ramaty, Microwave and hard x-ray bursts from
    solar flares, Solar Physics, VIII, 119, 1969.

Hudson, H. S., L. E. Peterson, and D. A. Schwartz, The hard solar
    x-ray spectrum observed from the third orbiting solar
    observatory, Astrophys. J., 157, 389, 1969.

Karzas, W. J. and R. Latter, Electron radiative transition in a
    coulomb field, The Astrophysical Journal Supplement series,
    VI, 55, p. 167, 1961.

Kreplin, R. W., T. A. Chubb, and H. Friedman, X-ray and lyman-alpha
    emission from the sun as measured from the NRL SR-1 satel-
    lite, J. Geophys. Res., 67, 2231, 1962.

Kundu, M. R., Solar Radio Astronomy, University of Michigan Radio
    Observatory Report No. 64-4, Ann Arbor, 1964.

Mandelstam, S. L., Usp. Fiz. Nauk., 46, 145, 1952.

Mandelstam, S. L., X-ray emission of the sun, Space Sci. Rev., 4,
    587, 1965.

Meekins, J. F., R. W. Kreplin, T. A. Chubb, and H. Friedman, X-ray
    line and continuum spectra of solar flares from 0.5 to 8.5
    angstroms, Science, 162, 891, 1968.

Neupert, W. M., W. Gates, M. Swartz, and R. Young, Observation of
    the solar flare x-ray emission line spectrum of iron from
    1.3 to 20 Å, Astrophys. J., 149, 179, 1967.

REFERENCES (Cont.)

Potemra, T. A. and A. J. Zmuda, The effect of dumped particles on the nighttime D-region, paper presented at the Spring Meeting of URSI, Washington, D. C., 21-24 April 1969.

Reidy, W. P., G. S. Vaiana, T. Zehnpfennig, and R. Giacconi, Study of x-ray images of the sun at solar minimum, Astrophys. J., 151, 333, 1968.

Seaton, M. J., The spectrum of the solar corona, Planet. Space Sci., 12, 55, 1964.

Underwood, J. H., Solar x-rays, Science, 159, 383, 1968.

Vaiana, G. S., W. P. Reidy, T. Zehnpfennig, L. Van Speybroeck, and R. Giacconi, X-ray structure of the sun during the important 1N flare of 8 June 1968, Science, 161, 564, 1968.

Van Allen, J. A., The solar x-ray flare of July 7, 1966, J. Geophys. Res., 72, 5903, 1967.

Van Allen, J. A. and S. M. Krimigis, Impulsive emission of ~ 40 kev electrons from the sun, J. Geophys. Res., 70, 5735, 1965.

Van Allen, J. A. and C. D. Wende, The solar flare of 8 July 1968, J. Geophys. Res., 74, 3046, 1969.

Walker, A. B. C. and H. Rugge, Space Research, 9, 1969 (to be published).

Wende, C. D., The correlation of solar microwave and soft x-ray radiation, 1. The solar cycle and slowly varying components, J. Geophys. Res., 74, 4649, 1969a.

Wende, C. D., The correlation of solar microwave and soft x-ray radiation, 2. The burst component, J. Geophys. Res., 74, 64715, 1969b.

Acknowledgements

The authors wish to express their thanks to Dr. J. A. Van Allen of the University of Iowa for providing the x-ray data from the Injuns, Mariner, and Explorers spacecraft, and for helpful comments regarding the results. Thanks are also due to Drs. W. Neupert and J. Underwood of the Goddard Space Flight Center for making original pictures and figures available. Drs. G. Vaiana and T. Zehnpfennig of American Science and Engineering supplied original pictures of their x-ray photographs.

This research was supported in part by the National Aeronautics and Space Administration under Task I of Contract NOw 62-0604-c.

RADIO-ASTRONOMICAL OBSERVATIONS

MICROWAVE SPECTRAL OBSERVATIONS OF CORONAL CONDENSATIONS

R. M. Straka

Radio Astronomy Branch, Ionospheric Physics Laboratory

Air Force Cambridge Research Laboratories, Bedford, Mass.

A. INTRODUCTION

The variation of solar radio flux with solar cycle, known as the Slowly Varying Component, has been attributed to the supplementary flux from dense condensations in the corona over solar disk active regions. Condensations have kinetic temperatures T_e (1 to 4×10^6 °K) which differ little from the surrounding 'normal' coronal temperatures. Their densities, $N \sim 10^9$ cm^{-3}, however, are about 5-10 times the normal density. These increased densities N become important as the bremsstrahlung brightness temperature T_b, given by $T_b = \int_0^{\tau_{max}} T_e\, e^{-\tau}\, d\tau$, is dependent on the optical thickness τ produced by collisional absorption, $\tau_{o,e}^{coll} \propto \frac{1}{\omega^2} \int \frac{N^2 dl}{T^{3/2}}$. Condensations also are source regions for bremsstrahlung X-ray emission. Rocket-borne X-ray photographs of the sun obtained by Friedman (1961) showed that the locations of intense X-ray sources were in very close agreement with the positions of 9.1 cm wavelength radio condensations as obtained by Stanford. Source sizes of the condensations are only several minutes of arc and therefore special high resolution radio techniques must be employed to isolate them for study. These techniques include (a) high resolution two-dimensional mapping with pencil beam interferometers (Christiansen et al., 1960; Swarup et al., 1963; Tsuchiya and Nagane, 1967); (b) high resolution one-dimensional drift scans (Covington, 1963; Tanaka and Steinberg, 1964); (c) total sun fluxes, where only the region of interest is on the solar disk (Abbasou et al., 1968); (d) and, with broad beam observations at solar eclipses using the moon's obscuration edge to obtain resolution tens of seconds in arc (Christiansen et al., 1949; Hachenberg et al., 1963; Castelli et al., 1963; Poumeyrol, 1967) to mention only a few. However, for all the observations made, the number of regions which still make up our total

knowledge of the coronal radio condensations are few indeed. The
large variation of sizes and configurations necessitate that many
more measurements be made to obtain the physical conditions in condensations.

B. OBSERVATIONS

To obtain spectral radio data of the condensation regions responsible for the S.V.C., the Air Force Cambridge Research Laboratories operated a multiwavelength 8-ft diameter radio telescope at
the 20 May 1966 Annular Solar Eclipse in Greece; and, at the 12
November 1966 Total Solar Eclipse in Peru. A special antenna feed
allowed the wavelengths of 3.4, 6.0, 11.1 and 21.2 cm radiation to
be simultaneously monitored. The wavelengths were selected to coincide exactly with the AFCRL Sagamore Hill Radio Observatory wavelengths, thus allowing whole-sun flux calibrations to within ±10%
of absolute for the eclipse measurements. The equipment has been
previously described by Castelli and Straka (1966). Local circumstances of these two eclipses are described in Table I.

TABLE I - ECLIPSE EPHEMERIS

	20 May 1966	12 November 1966
1st Contact	08:04:37 U.T.	11:58:00 U.T.
Maximum Phase	09:31:24	13:02:34
4th Contact	11:04:52	14:15:32
Magnitude of Eclipse	0.999	1.018

C. RESULTS

C.1 Analysis

C.1.a Slope Determination. In order to determine the changes in
total flux resulting from the covering and uncovering of the condensations during the eclipse, the derivative $\frac{d\Phi_\lambda}{dt}$ of the total flux
curve $\Phi_\lambda(t)$ was obtained. Derivative values were calculated on a
minute by minute basis to smooth this 'slope' curve. Time shifts
were avoided by using the Runge-Cutta method for obtaining the derivative:

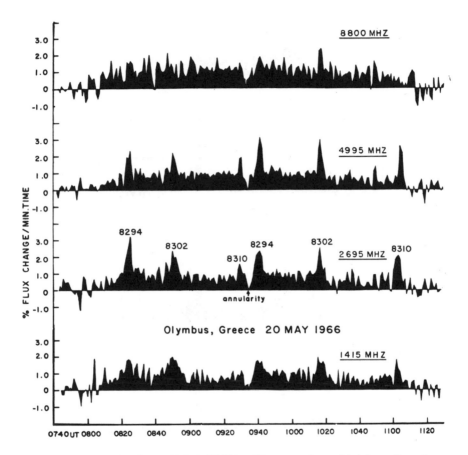

Figure 1. 20 MAY 1966 SLOPE CURVES: The percentage of total sun flux change per minute of time is plotted as a function of Universal Time. Both covering and uncovering phases have been arranged to show positive slope values. Mc Math Plage Numbers indicate the region wherein the radio sources are located.

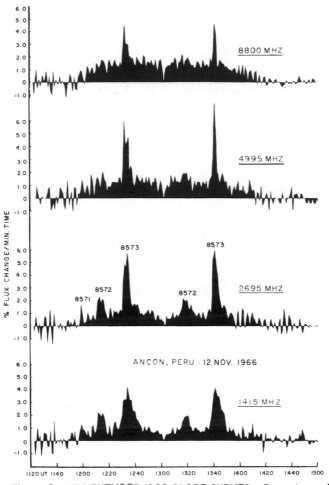

Figure 2. 12 NOVEMBER 1966 SLOPE CURVES: Percentage of
total sun flux change per minute of time as a function of time.
The radio sources are assigned the Mc Math numbers of their
associated plage regions.

$$\frac{d\left(_n\Phi_\lambda\right)}{dt} = \frac{\left(_{-2}\Phi_\lambda\right) - 8\left(_{-1}\Phi_\lambda\right) + 8\left(_{+1}\Phi_\lambda\right) - \left(_{+2}\Phi_\lambda\right)}{12h}$$

where $\dfrac{d\left(_n\Phi_\lambda\right)}{dt}$ = slope of flux curve at instant of time t_n

$\mp_K \Phi_\lambda$ = flux value at $\mp k$ minutes from t_n

h   = interval in minutes over which slope determined (h = 1 minute in our case)

Changes of 1 flux unit (1×10^{-22} w/M^2/Hz) were detectable at all wavelengths. Figures 1 and 2 show the slope curves obtained from the 20 May and the 12 November eclipses respectively. The ordinate values are given in percentage flux change (of total sun flux) per minute of time of the eclipsing action. Negative slope values are obtained during the first half of the eclipse, but are shown here as positive values to make the radio regions graphically more obvious. Peaks in the slope values coincide with the covering and uncovering of radio coronal condensations. The McMath numbers of the plage regions associated with the condensations are given. To obtain the flux value F associated with each of the coronal condensations the following summation,

$$F = \sum_n \left(\frac{d\Phi(t)}{dt}\right)_n \cdot \Delta t$$

is performed over the duration of the maximum in slope. However, the total sun flux $\Phi_{total} = \Phi_{background\ sun} + \Phi_{condensations}$. It therefore becomes necessary to subtract the slope contribution of the background (quiet) sun from the eclipse slope curve $\dfrac{d\Phi(t)}{dt}$ to obtain the true flux value of the condensation.

C.1.b Quiet Sun Model. In the past, several methods have been employed to generate the quiet sun model (e.g. Christiansen et. al., 1949; Eriksen et. al., 1955; Covington et. al., 1955; and Drago et. al., 1963). These models ranged from eclipsing a simple homogeneous solar disk to complex mathematical models which could take into account the limb brightening and polar darkening observed at radio wavelengths. The method AFCRL used takes these latter conditions into account, but additionally provides for the possible East-West asymmetry of the quiet sun as noted by Roosen and Goh (1967). To obtain this model each of the matrix values of the Stanford 9.1 cm maps were averaged over 44 days in July, August, September 1964 when sunspot numbers were zero. Radio fluxs from the sun were at their lowest values for the solar cycle during the latter half of July 1964. By averaging each of the matrix points, changes due to weak radio regions

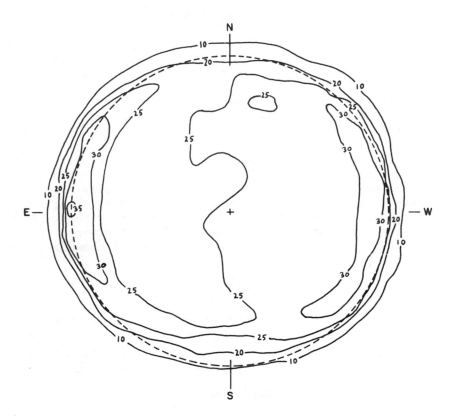

Figure 3. MODEL OF THE QUIET SUN: Brightness contour maps constructed
     from averaged matrix values of the 9.1cm Stanford Maps during a 44 day
     quiet period in July, August, and September of 1964. This map was
     artificially eclipsed to provide the slope curves of the quiet sun background.
     Contour lines are in units of 1000°k .

not associated with sunspots would be smoothed out. Figure 3 shows
the resultant quiet sun 9.1 cm map. This map was then artificially
eclipsed using the same moon-sun geometry as was present during the
two 1966 eclipses. To match the 9.1 cm model to the four wavelengths used by AFCRL, the 9.1 cm values were scaled spectrally in
the ratio of their total sun flux values. Table II lists the total
sun flux values used:

TABLE II - TOTAL SUN FLUXES
(in units of 10^{-22} w/M^2/Hz)

Wavelength	20 May 1966	12 November 1966	Quiet Sun Model
3.4 cm	293	298	233
6.0 cm	147	161	100
11.1 cm	105	132	68
21.2 cm	69	88	37

Quiet sun flux values were obtained by averaging the Nagoya 1000 Mhz,
2000 Mhz, 3750 MHz, and 9400 MHz, and the Ottawa 2800 MHz flux values
for the 44 day quiet period mentioned above, and appropriately scaling
to our frequencies. One uncertainty is that of the degree of asymmetry of the quiet sun at 3.2 cm, etc., compared to that at 9.1 cm.
This difference, however, must be small when compared to other uncertainties in the data as a good match of the artificially obtained
flux curve to the actual eclipse flux curve was obtained for all the
AFCRL wavelengths.

C.1.c Maximum Phase of the Radio Eclipse. Propagation of solar
radio waves can occur only above the zero refractive-index (n = 0)
level in the solar atmosphere. Below this level the plasma frequency f_o is greater than the wave frequency f, so that n becomes
imaginary and the radio wave is strongly damped. Because of the
decrease in electron density N outward from the sun, the longer-
wavelength radiation has an effective emission altitude higher in the
solar atmosphere. Therefore, during the maximum phase of an eclipse
the longer wavelength radio sun is less obscured and gives a higher
residual flux level. The residual flux levels measured during the
two 1966 eclipses are given in Table III. Included in this table
are the times when maximum obscurations of the radio sun occurred.
These times may differ from the optical maximum phase times due to
possible asymmetry of the radio quiet sun background, or more likely,
from the presence of active regions near the solar limbs.

TABLE III - MAXIMUM OBSCURATION OF RADIO SUN

	20 May 1966 Residual	20 May 1966 Max Phase	12 Nov. 1966 Residual	12 Nov. 1966 Max Phase
Optical	(0.999 mag)	0931:24 U.T.	(1.018)	1302:24 U.T.
3.2 cm	7.1 %	0932:50 ±15$_s$	5.7 %	1302:33 ±15$_s$
6.0 cm	11.9 %	0933:26 ±15$_s$	9.0 %	1302:28 ±15$_s$
11.1 cm	16.0 %	0934:05 ±15$_s$	11.5 %	1302:17 ±15$_s$
21.2 cm	18.5 %	0934:28 ±15$_s$	14.1 %	1301:33 ±15$_s$

C.1.d Flux Values of Condensations. The fluxes obtained from each
of the condensation regions are tabulated in Table IV, and plotted
in Figure 4. All sources (except that associated with 8573) had
maximum fluxes less than 10 f.u., which according to the Swarup
(1963) classification scheme, means these sources can be considered
as moderate to weak in intensity. The classification is that at
9.1 cm wavelength, 3-8 f.u. = weak source, 8-15 f.u. = moderate source,
and >15 f.u. = strong source (when the source is within three days
of CMP). The location of the condensation regions are shown in
Figures 5 and 6. Shown on Fraunhofer Institute optical maps are the
moon's limb positions for times when the 11.1 cm wavelength slope
maximums occurred. Intersections of these limb positions show the
locations of radio sources overlying the sunspot and bright plage
regions.

C.1.e Brightness Temperatures and Source Sizes. The brightness
temperature of the condensation regions are determined from:

$$T_b = \frac{F\lambda^2}{2K\Omega}$$

where Ω is the solid angular extent of the region in steradians, F
is the region flux at wavelength λ , and $K = 1.38 \times 10^{-23}$. It is
possible to determine the angular extent Ω of the condensations by
noting the half-power time duration of the flux slope curves and
multiplying by the velocity factor of the moon's limb. For the 20
May eclipse this velocity factor was 21" arc/min of time; and 30"
arc/min of time for the 12 November eclipse. One second of arc then
corresponded to a distance D of 740 kilometers and 717 kilometers,
respectively. Angular sizes, in minutes of arc, for all the sources
during both eclipses are presented in Figure 7. The increase of size
with λ >10 cm is in good agreement with past observations of other
researchers (e.g. shown by the solid curve representing Swarup et.
al., (1963) data). Although the trend as a function of frequency is
the same, the sizes in the Swarup curve are larger. Sources used in
his study all had associated sunspot regions with areas > 500 mil-

TABLE IV

McMATH PLACE	PLACE LOCATION	SUNSPOTS ZURICH CLASS	SUNSPOT AREA 10⁻⁶ of DISK	MAG. CLASS. MT. WILSON	FLUX (10^{-22}w/m^2/cps)				BRIGHTNESS TEMP (x 10^6 °K)			
					21.2cm	11.1cm	6.0cm	3.2cm	21.2cm	11.1cm	6.0cm	3.2cm
20 May 1966												
Athens												
8294	23N 58W	D13	140	β	3.9	7.0	7.2	9.6	2.39	2.42	1.23	0.27
8300	31N 22W	A2	10		0.8	0.8	0.7	3.5	4.76	2.07	0.72	0.53
8301	23N 12W				2.8	1.5	1.8	2.1	4.00	2.16	0.55	0.22
8302	22S 5E	C14	50	β$_p$	4.3	5.2	6.3	8.8	2.70	2.44	1.24	0.56
8310	17N 71E	A1 D12	100	α$_p$	2.2	6.4	6.0	7.5	3.74	1.34	1.20	0.39
12 Nov 1966												
Sac Peak				Photo. field gauss								
8571	18N 82W	J4			1.4	1.7	1.3	1.8	0.60	0.32	0.18	0.02
8572	28N 18W	C15	234	β$_p$ 1600-2000 α$_p$ 1100-1500	4.6	6.5	5.5	6.3	1.54	0.75	0.32	0.30
8573	13N 15E	H16	340	β$_p$ 2100-2500	13.4	16.4	16.9	13.4	4.19	3.42	1.02	0.26
		H10	478	α$_p$ 2600-3000	7.7	13.1	11.8	14.3	4.70	4.00	2.32	0.91
		H16 & H10	(uncovery)		19.3	26.5	20.0	18.8	3.45	2.87	2.32	0.70
Mound Prom.	(35N)				0.2	0.7	0.7	1.8	1.77	1.78	0.52	0.18

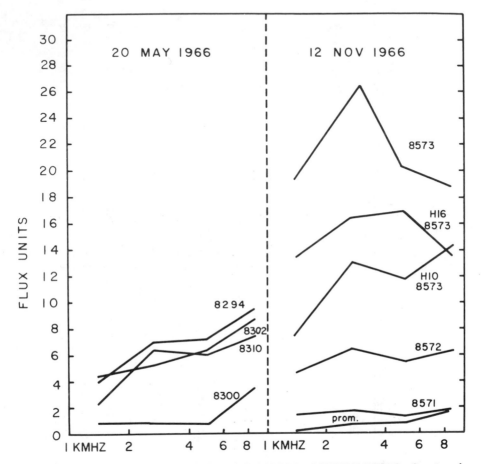

Figure 4. FLUX SPECTRA OF THE RADIO CORONAL CONDENSATIONS: Spectra of
the radio sources during the 20 May and 12 November eclipses of 1966. One flux unit
equals 1×10^{-22} w/m^2 / MHz. Three curves are shown for the source in Mc Math
region 8573. Two are for spot groups H10 and H16 when they could be resolved
during the covering. The third and uppermost curve represents both spot groups
simultaneously being uncovered. The west limb (35 N) mound prominence
spectrum is also shown.

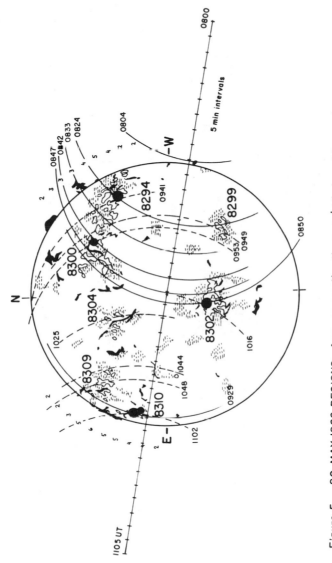

Figure 5.   20 MAY 1966 REGIONS: Arcs representing the moon's limb position at times when maxima were noted on the 11.1 cm slope curves are plotted on the Fraunhofer Institute Map. Center of the moon positions are given at 5 minute time intervals.

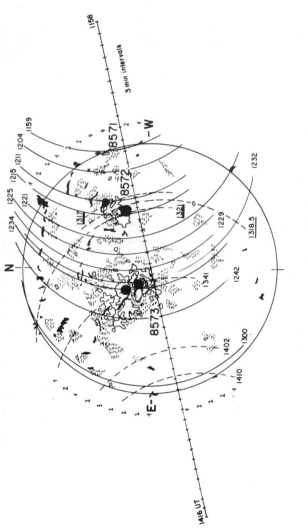

Figure 6.   12 NOVEMBER 1966 REGIONS: Arcs representing the moon's limb position at
times when maxima were noted on the 11.1 cm slope curves are plotted on the Fraunhofer
Institute Map. Center of the moon positions are given at 3 minute time intervals.

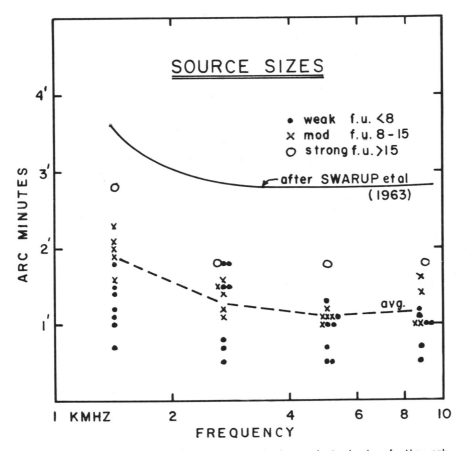

Figure 7. SOURCE SIZES: The angular extension in arc minutes is given for the weak, moderate, and strong radio sources visible during the 20 May and 12 November 1966 eclipses. Note the large sizes as a function of frequency for region 8573 (denoted by circles). The dashed curve represents the average size of the moderate sources.

lionths of the disk. None of the 20 May nor 12 November spot regions
were this large. The exception was the source in plage 8573 with
an associated sunspot area of near 478 millionths. Note the larger
sizes of this source in Figure 7 as shown by circles. The other
sources had associated sunspot areas of ≪500 millionths. In general,
the more intense sources have larger sizes at all wavelengths. For
the moderate intensity sources, sizes and dispersion of sizes are
shown to be least near 6 cm wavelength. The largest sizes and
greatest dispersion are noted at 21 cm wavelength, with similar but
lesser effects at 3.2 cm. The usual consensus is that there is close
association of the 6-10 cm wavelength source sizes with underlying
sunspot extensions, and with plage area for the longer wavelength 20
cm emission sources. Zheleznyakov (1964) suggests also that at
shorter wavelengths, say $\gtrsim$ 3 cm, one again starts to have closer
association with the thermal plage area below. It must be noted
that in obtaining half-power widths of sources by the eclipse
method, complications arise when two sources are simultaneously
eclipsed by the moon's limb. A case in point occurred during the
uncovering of the H16 and H10 sunspots in McMath Plage region 8573
(1341 U.T. in Figure 6) when the long northeasterly extension of
the plage region was simultaneously uncovered.

Brightness temperatures T_b were calculated using the sizes of
the sources as obtained above and by assuming circular symmetry for
these sources. The values appear in Table IV. Considering that
the brightness temperature is related to the kinetic temperature of
the condensation region by

$$T_b = \frac{T_e}{2}\left[(1 - e^{-\tau o}) + (1 - e^{-\tau e})\right]$$

then, in the long cm wavelength range where the optical depth $\tau_{o,e}$
becomes one or greater, T_b becomes high and approaches the value of
T_e. Brightness temperatures of the order of 4.7 x 10^{6}°K were
measured at the longest observational wavelength of 21 cm. Electron
temperatures of this order (4 x 10^{6}°K) have been reported by Swarup
et. al., (1963) and optically by Widing (1966) and Neupert (1967).

C.2   Region Characteristics

C.2.a   **20 May 1966.** Region 8294 The radio source in this region
was over a growing sunspot of class D13 (Zurich class at CMP) located at 18N 52W. This spot was listed by Mt. Wilson on 19 May (no
observations on 20 May) as a β magnetic class spot. The plage region
surrounding the spot is an old region on its third or fourth rotation. On its passage across the disk the region was quiet in flare
production; the maximum number of subflares occurring on 19 May.

Region 8302 This source was associated with the C14 spot group
(20S 5E) which was born on 18 May and was still growing at the time

of the eclipse (i.e., area from 46 millionths of the disk on the
20th to 155 millionths on the 21st.) Subflare activity in the area
was reported on eclipse day. It continued producing subflares during
its passage across the disk. An unexpected feature occurred during
the eclipse, in that no change was noted at 3.2 cm for this region
during the covering at 0850 U.T., whereas, on the uncovering a significant change (8.8 flux units) occurred. Looking at the original
records one finds, if anything, a slight increase in flux during
the covering. An examination of 3.16 cm records taken only a few
miles from the AFCRL site, and loaned to us by the Arcetri Observatory, also shows no flux change during the covering. Unfortunately,
during the uncovering they were in a calibration mode. However,
other site observations of this same region by the Arcetri group
(Drago and Noci, 1969) show a flux at 3.1 cm of 6.7 ± 2.2 flux
units where we measure 8.8 flux units. At 9.1 cm the Arcetri group
measures 5.1 ± 0.2 f.u. which agrees closely with our 5.2 f.u. at
11.1 cm wavelength. Fürstenberg et. al., (1967) report in their
Figure 3 values of 5.5 f.u. at 3.28 cm, 3.9 f.u. at 10.3 cm, 2.5
f.u. at 15 cm, and 2.2 f.u. at 20.1 cm. These are somewhat lower
than both the AFCRL and Arcetri values. Apustinskiy et. al., (1968)
give the following flux values: 2.3 f.u. at 0.8 cm, 2.0 f.u. at
1.6 cm, 2.3 f.u. at 3.3 cm, and 7.0 f.u. at 10 cm. Observations for
this region have also been reported by Poumeyrol (1967) at 32 cm
wavelength giving a flux of 1.94. The region was shown to be an
important source of X-ray flux by the Solrad 8 satellite observations
of Landini et. al., (1966).

Region 8310   The A1 and D12 spot groups located 17N 71E in plage
8310 are simultaneously covered and uncovered together so that
separation between the two components was unattainable. Plage region
8310 is on its first rotation and produced a number of subflares, Imp
I flares, and on 28 May produced a 2B flare. Intense 5303 Å
coronal lines (121-160 x 10^{-6} of the sun's brightness) were seen near
this region on the east limb.

Region 8300   The source in this plage was located over an A2
sunspot at N28 W26. It was eclipsed time-wise first at 3.2 centi-
meters, then 6.0 and 11.1 cm, and after a shift of about 50" arc
eastward finally the 21.2 cm wavelength source. This is opposite to
what would be expected for different wavelength emission centers arranged radially in height over the spot. Normally in this case, the
21.1 cm radiation would be eclipsed first, then the shorter wavelengths. It appears that the 21.1 cm source was aligning itself
over the center of gravity of the plage rather than of the spot.
Region 8302 was also being covered simultaneously and may in part be
responsible for the source position shift. However, during the uncovering, plage 8302 could not contribute to a shift, although perhaps 8301 did.

Other Regions Small but definite changes in flux were noted
when plage regions 8311, 8304, and 8309 were eclipsed. The radio
source over plage 8304 contributed about 1.8 percent of the total
flux at 9.1 cm (as scaled from the Stanford maps in the ESSA Solar
Geophysical Data Bulletin). The small source associated with plage
8309 had 4.1 flux units at 3.2 cm, 1.5 f.u. at 6.0 cm, 0.7 f.u. at
11.1 cm, but was not detectable at 21.2 cm. Effects from a limb
prominence at   ~15S were measured just prior to first contact.
Another west limb prominence(at N 35) gave flux changes, however,
the eastern portion of Plage 8394 may in part have been responsible
for the changes.

C.2.b 12 November 1966. Region 8573 Of both eclipses, this radio
condensation was the largest in extent and had the highest flux
levels. The plage region was continually in growth throughout the
eclipse and was a third and fourth rotation region which was previously McMath numbers 8545 and 8546. The H10 spot group and H16
spot group were of Mt. Wilson classification α_p (maximum field
of 2600-3000 gauss), and β_p (maximum field of 2100-2500 gauss) respectively. Many flares were associated with this complex region
including the importance 2n flare of 7 November. On 12 November,
Type III radio burst activity and a 200 MHz noise storm from 0816 -
1206 U.T. were associated with this region. MacKenzie University
(Kaufmann, 1968) measured the radio condensation region with a
polarimeter at 4.28 cm. He obtained a flux density of 22 flux units
for the composite region and, more importantly, found that the following bipolar H16 group had polarized radiation with a brightness
$T_b \approx 0.7 \times 10^{6\,o}K$. The leading unipolar H10 spot group, on the
other hand, had an unpolarized radiation with brightness temperature
$T_b \approx 1.2 \times 10^{6\,o}K$. Polarization for the whole region was about 50
percent left-handed. This then is a region (from both eclipses)
where one begins to see the magnetic influence on the flux spectrum
of the condensation.

Region 8572 Plage 8572 was on its first rotation and rapidly
growing on the day of the eclipse. The small leading spot group was
magnetically classed by Mt. Wilson as α_p (maximum field of 1100-
1500 gauss) and the larger trailing spot group as β_p (maximum field
of 1600-2000 gauss.) Both spots were resolved at 3.2 and 6.0 cm
during the covering phase of the eclipse, with a suggestive but unresolved result at the two longer wavelengths. During the uncovering,
only the 3.2 cm sources were resolved, with both spot groups being
treated as a single region at the other wavelengths.

Region 8571 This first-rotation plage region was just about to
rotate off the disk on eclipse day. On 8 November there had been a
B6 spot in the plage, but it was gone by the 9th. Coronal lines at
5303 Å were moderate over this region. X-ray photographs taken by
Dr. Underwood of NASA (1967) revealed a bright x-ray region in the
vicinity of this source.

D. DISCUSSION

A number of different interesting solar condensation regions
were present during the 1966 solar eclipses. It was significant that
all the regions with fluxes less than 10 flux units tended to show
a spectrum characteristic of thermal bremsstrahlung emission. The
peak in the spectrum near 6 cm wavelength was not noted as reported,
for example, by Molchanov (1964) and Swarup et. al. (1963). Regions
in those studies had fluxes well above 10 f.u. and could be categor-
ized as strong sources. Kakinuma and Swarup (1962) and Zheleznyakov
(1962) independently proposed that contributions from gyro-resonance
radiation at $\omega \simeq S\omega_H$ (mainly at S = 2, 3) could add appreciably to
the electro-ion collisional thermal radiation from the condensation
region. With the addition of gyro-resonance absorption mechanisms,
not only could the peak in the emission spectrum be accounted for but,
as is observed, that the degree of polarization is greater at 3 cm
than at 10 cm. At wavelengths above the peak wavelength (λ m),
bremsstrahlung thermal emission again predominates the spectrum as
the increased optical thickness leads to considerable absorption of
the gyro-radiation. At the short wavelength end of λ_m a decrease
in the gyro component also occurs due to the transition of the
   $\omega \simeq 2\omega_H$ and $\omega \simeq 3\omega_m$ radiating levels from the high temperature
corona to the low temperature chromosphere, the heights being depen-
dent on the sunspot magnetic field strengths. Referring back to
Figure 4, those regions which are not well developed, and/or associated with weak sunspot fields, have predominantly bremsstrahlung
spectrum. Peaking in the radio spectrum does show up however in the
bipolar H16 spot region (in plage 8573) where sufficient magnetic
field strengths and configurations exist to make the effect of gyro-
resonance radiation a governing factor of the spectrum. Recalling
that Kaufmann (1968) measured about 50 percent left-circular polarization of this region points to the gyro-resonance effect on the
polarization of sources. This polarized radiation could be explained
if the extraordinary brightness temperature T_{be} is greater from the
   $\omega \simeq 3\omega_H$ level than the ordinary T_{bo} from the $\omega \simeq 2\omega_H$ level
in the deeper and cooler layers. Since the polarization
$\rho = \dfrac{T_{be} - T_{bo}}{T_{be} + T_{bo}}$ a large difference in T_{be} and T_{bo} gives a high value
of ρ. At increasing λ the $\omega \simeq 2\omega_H$ and $\omega \simeq 3\omega_H$ levels
both shift higher into the solar atmosphere where their difference
becomes smaller, causing a resultant decrease in polarization. Regarding polarization measurements, Zlotnik (1968) extended the caution
that unless the polarization is measured where the magnetic fields are
effective in the region, rather than averaged over the entire source
where sectors of unpolarized bremsstrahlung can contribute, the obtained degree of polarization would be less than actually exists.
The difference in brightness distribution over the source has been
pointed out by Kundu (1959). He observed at 3 cm wavelength a bright
core for the polarized component with a surrounding halo from the un-

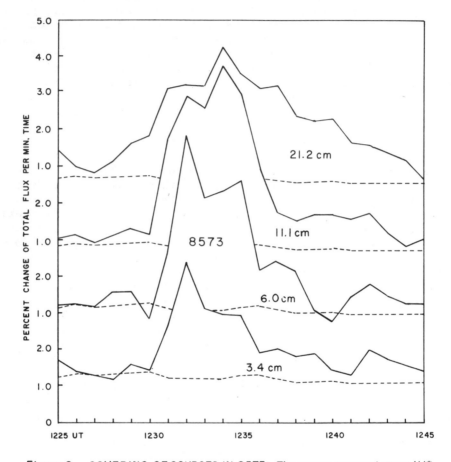

Figure 8.   COVERING OF SOURCES IN 8573: The source over spot group H 10
peaks first during the covering, then, the second peak corresponds to the H 16
spot group source. Zero percent indices are displaced for the upper three
curves. Dashed curves represent the slope values of the quiet sun background.

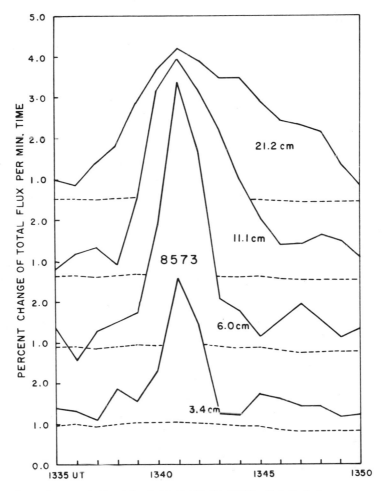

Figure 9. UNCOVERING OF SOURCES IN 8573: Detailed slope curves during the uncovering of H10 and H16 spot groups in 8573. Both groups are uncovered simultaneously.

polarized component. Unpolarized radiation measurements would show
little of this bright core since the high value of the extraordinary component primarily responsible for polarization would be
compensated by the reduced ordinary component (Zlotnik, 1968).
Figures 8 and 9 show the AFCRL multiwavelength observations of
region 8573 in detail. During the eclipse both spot regions were
resolved. Note the different spectrum from both regions. Extension
of the source width at 21.2 cm can also be seen. Figure 9 shows
both sources were uncovered simultaneously. During the uncovering
phase, more of the north-east extension of plage 8573 is aligned
with the moon's limb, so that the plage is uncovered in less time
than it took to cover it. Note then the differences in the 21.2
cm source widths in Figures 8 and 9 that coincide with the asymmetry
of the plage eclipsing.

In addition to the spectrum of the source, reliable height
distributions of brightness temperatures must be obtained before an
electron density distribution in the condensation can be determined.
Christiansen et. al. (1960) employed measured brightness temperatures
and associated heights in the relations:

$$T_b = \int_0^{\tau max} T_e \, e^{-\tau} d\tau \quad \text{and} \quad \bar{\rho} = \frac{\int_0^{\tau max} \rho \, dT_b}{\int_0^{\tau max} dT_b}$$

to analytically obtain the density distribution $N(\rho)$. Heights for
this study were determined from the rate of travel of the regions
across the disk as obtained by high resolution interferometers. A
method that is employed during eclipses to obtain height of the
condensation regions as a function of wavelength λ is to note time
differences between the eclipsing of the optical spot centers and
the radio sources. For example, a source at the west limb would be
covered first at the longest λ, with successively shorter wavelengths being covered at later times. However, if the center of
gravity of the decimeter radiation shifts from over the spot to
over the larger plage area, then a non-radial height distribution
as a function of wavelength occurs. Measurements of time differences
in covering and uncovering would then no longer be directly related
to source heights at the different wavelengths. A case in point is
the source associated with plage 8300 of the 20 May eclipse. During
the uncovering (covering effected by plage 8302) of the A2 spot
region at longitude 26W the 3.2 cm radiation was uncovered at
0952:30 ± 30s, 6.0 and 11.1 cm later at 0953:30 ± 30s, and finally
21.2 cm at 0954:00 ± 30s. In the western quadrant the 21.2 cm source
should have been uncovered first, then at the shorter wavelengths.
Apparently, an eastward shift of the center of gravity occurred as
the decimeter radiation aligned itself with the eastward extension
of the plage region (Figure 5). The other regions (8294, 8302, and
8310) however, had their wavelength displacements in the directions
expected from a radial distributions, i.e., for 8294 the 21.1 cm
source is west of the 3.2 cm source, for 8302 very little displace-

ment occurs because of the near central meridian location, and for
8310 the 21.1 cm source is east of the 3.2 cm source. Normal displacements occurred during the 12 November eclipse, with the exception of the covering of the C3 spot at 39W in the plage region
8572. Here again, the 21.1 cm source was east of the 3.1 cm source.
The spot was on the western edge of the plage region, as shown in
Figure 6. During the uncovering, the C3 and D13 spots were resolved
only at 3.2 cm. This is indicated by the doubly underlined times
1317 and 1321 U.T. in Figure 6. Other wavelengths had positions
averaged somewhat north and east of the D13 spot as shown at 1318.5
UT.

A paper by Zlotnik (1968a) shows how displacements of the radiation center from the true source center can occur. He attributes
the shifts to the relative positions of the emitting gyro-resonance
levels with respect to the line of propagation. Figure 7 of his
paper shows shifts in the radiation center for both the $\underline{e}$ and $\underline{o}$
modes of 9 cm wavelength radiation as a function of viewing angle
Ψ . Shown are source displacements of 10,000 km and 25,000 km
from the spot center for Ψ = 30 and 60° respectively. Zlotnik indicates a "restructuring" of the brightness distribution over the
source as Ψ changes is thus responsible for the source displacement
and not the geometric localization of source height above the photosphere. Another effect of longitudinal position of the condensations
is given by Swarup (1963). He shows that the fluxes of sources
(both strong and moderate sources grouped together) at 21 cm closely
follow a cosine law with heliographic longitude, 9.1 cm sources the
least change in flux with longitude, and again at 3 cm almost a
cosine effect. Additionally, it is shown that the strong radio
regions at 9.1 cm wavelength have the greatest change with solar longitude, approaching that of a cosine variation. Moderate regions have
a lesser change and weak regions the least. If the source is optical-
ly thick ($\tau = \mu h \gtrsim 1$, $\tau = \mu d \gtrsim 1$)   with a sufficiently
high value of the ratio d/h then the source flux will have a cosine
Ψ dependence, as for example 21 cm wavelength sources. τ is the
optical depth, μ the absorption coefficient, h the altitudinal ex-
tent, and d the longitudinal extension. Optically thin sources ($\tau = \mu h \ll 1$, $\tau = \mu d \ll 1$) have a near constant flux with heliographic
longitude. From Swarup's wavelength dependence relationships men-
tioned above it appears that the 3.2 cm source may be optically thick
as it too approaches the cos Ψ variation.

Zheleznyakov (1962, 1964) has shown that the gyro-magnetic intensity is very sensitive to the angle α between the propagation
direction and the field H_o. As the source approaches the limb the
gyro component becomes less of a contribution to the total source
flux and free-free bremsstrahlung remains the dominant emission
mechanism. The opacity changes in the $\omega = S\omega_H$ layers responsible
for this effect, as well as the field geometries of thermal gyro-
magnetic radio sources, are shown by Lantos (1968). The decrease

in gyro emission (near the spectral maximum λ_m) with heliographic
longitude could account for the greater longitudinal effect on
strong sources, which have gyro emission, compared to that of the
weaker predominantly bremsstrahlung sources. The high longitude
sources in regions 8310 and 8294 of the 20 May eclipse appear to
have been influenced by this effect. Another aspect that Zheleznyakov
points out, is that with the decline of the spot region and disappearance of magnetic fields the emission from the condensation region
becomes of the bremsstrahlung type, irrespective of what it had been.
From the Sac Peak data (Carrigan, 1966 and 1966a) the spot history
during the two eclipses were given as:

20 May 1966				History on Disk (ESSA Bulletin)
Spot in McMath	8294	growing		b ⌐ ℓ
	8302	growing		b ∧ ℓ
	8310	growing		ℓ ∧ d
	8300	constant		ℓ - d
12 Nov 1966				
	8573 H10	growing		b - d
	H16	constant		b - d
	8572 D13	growing		b ∧ ℓ
	C3	growing		b ⌐ ℓ

During the 20 May 1966 eclipse it is to be noted, in Table III,
that the maximum-phase of the radio eclipse occurs several minutes
later than the optical eclipse. The delay is also a function of
wavelength -- the longer the wavelength the greater the difference
in time between maximum phases. This apparent shift of the radio
sun center-of-gravity as a function of wavelength suggests that limb
activity was present on the east limb. Intense coronal line activity
at λ 5303 is indeed indicated at the east limb on the Fraunhofer Institute Map of Figure 5. If the X-ray sources are in the same altitude range as the cm radio condensations ($\sim$10,000 - 40,000 kms) then
a similar shift in the center of gravity of the x-ray sun could
occur. Minimum ionizing radiation would then not occur at the optical
minimum. The ionospheric physicist has been using the optical minimum to compare with the minimum electron density in the E and F1
regions to obtain the recombination coefficients α_E and α_{F1}, by

the equation

$$\alpha = \frac{1}{2N\Delta t}$$

N is the value of electron density when the production rate is minimum, and Δt the time lag or "sluggishness" between optical and ionospheric maximum phases of the eclipse. Clearly an error would result for an asymmetrical X-ray sun. It is therefore suggested that for recombination determinations of this type, the time of minimum ionizing flux, rather than the optical time of minimum, be used to obtain the delay Δt .   Even if the ionizing flux curves were not available, use of the radio flux minimum times would give better results than obtained with optical minimum times.

E.   SUMMARY

Analysis of 3.2, 6.0, 11.1, and 21.2 cm wavelength data from the two solar eclipses in 1966 on 20 May and 12 November indicate the following characteristics about the condensation sources of the S.V.C. of solar emission:

a.   Where the source flux was small ($<$ 10 flux units) the spectrum was predominantly of a bremsstrahlung thermal character.

b.   When the spot region became more developed and magnetic fields became a controlling factor, gyro emission at the $\omega = 2\omega_H$ and $\omega = 3\omega_H$ levels became important contributors to the flux from the condensation. Peaking in the spectrum around 6 cm wavelength then became apparent (e.g., source in McMath plage 8573 on 12 November).

c.   The 6 cm wavelength source sizes were the smallest and least variable compared to source sizes at longer and shorter wavelengths.

d.   The 21 cm sources were largest and had the greatest variation with source magnitude.

e.   Brightness temperatures as high as 4.7×10^{6} °K were measured in the condensation regions (at 21.2 cm) during both the 21 May and 12 Nov. 1966 eclipses.

f.   Maximum-phase of the radio eclipse (20 May 1966) occurred up to 3 minutes later than the optical maximum-phase due to east-limb activity shifting the center of gravity of the radio sun eastwards.

g.   It is suggested that for future determinations of the ionospheric recombination coefficients α_E and α_{F1} , where the time-delay method is employed, that use be made of the ionizing flux minimum time rather than the optical minimum time; and, that radio flux minimum times be used if ionizing flux curves are not available.

Even though characteristics of several more radio condensations have been obtained in these two eclipses, many more localized sources must be studied -- both weak and strong. High resolution studies on the continuing basis are needed to obtain accurate source positions, height distributions, and source sizes as a function of wavelength. Of vital importance is high resolution magnetic measurements in the vicinity of the sunspots and at the various condensation emission levels. With accurate heights and electron temperature determinations then correct density distributions in these dense regions in the solar atmosphere would be attainable.

F.   ACKNOWLEDGEMENTS

Mr. John P. Castelli of AFCRL was responsible for the initiation of these experiments and for the collection of the Greek eclipse data. His guidance and valuable efforts made the experiments the success they were. Messrs. C. Ferioli, E. Stewart, V. Remillard, W. Barron (AFCRL) and Sgt. D. Elwell (AWS) are to be thanked for their tireless efforts to maintain the field installations. Generous support to the expeditions was provided by Dr. J. Aarons (AFCRL), Professor M. Anastassiades (National Observatory of Athens), Eng. A. Giesecke (Institute Geofisico del Peru) and Dr. R. Waetjen (NASA Tracking Station, Ancon). A special note of gratitude is extended to the staff personnel of our host institutions in Greece and Peru for their valuable aid in establishing our sites.

REFERENCES

Abbasov, A.R., et. al.: 1968, Soviet. Astron. - A.J., 11, 1061.

Apushinskiy, G.P., et. al.: 1968, Vestn. Leningr. Univ., Ser. Kratkiye Nauchn. Soobshch., 7, 141.

Carrigan, A.L., and Oliver, N.J.: 1966, Geophysics and Space Data Bulletin, 3, 2nd Quarter.

Carrigan, A.L., and Oliver, N.J.: 1966a, Geophysics and Space Data Bulletin, 3, 4th Quarter.

Castelli, J.P. et. al.: 1963 ICARUS, 2, 317.

Castelli, J.P. and Straka, R.M.: 1966, Sky & Telesc., 32, 84.

Christiansen, W.N., et. al.: 1949, Australian J. Sci. Research, A2, 506.

Christiansen, W.N., et. al.: 1960, Ann. Astrophys., 22, 75.

Covington, A.E., et. al.: 1955, R.A.S.C. Jour., 49, 235.

Covington, A.E.: 1963, R.A.S.C. Jour., 57, 253.

Drago, F.G., et. al.: 1963, Ann. Astrophys., 27, 708.

Drago, F.G. and Noci, G.C.: 1969, Solar Phys. 7, 276.

Eriksen, G., et. al.: 1955, Astrophys. Norveg., Oslo, 5, 131.

Friedman, H.: 1961, Space Research II, ed. H.C. Van de Hulst, C. De Jager, and A.F. Moore, North-Holland Publishing Company, Amsterdam, 1021.

Fürstenberg, F., Keiser, J., and Priese, J.: 1967, Astr. Nachr., 290, 183.

Hachenberg, O., et. al.: 1963, Z. Astrophys., 58, 28.

Kakinuma, T. and Swarup, G.: 1962, Astrophys. J., 136, 975.

Kaufmann, P.: 1968, Solar Phys. 4, 58.

Kundu, M.R.: 1959, Paris Symposium on Radio Astronomy, Stanford Univ. Press, 222.

Landini, M., et. al.: 1966, Nature, 211, 393.

Lantos, P.: 1968, Ann. Astrophys., 31, 105.

Molchanov, A.P.: 1964, Izv. Astron. Obs. Pulkov., 24, 38.

Neupert, W.M.: 1967, Solar Phys. 2, 294.

Poumeyrol, F.: 1967, Ann. Astrophys., 30, 553.

Roosen, J. and Goh, T.: 1967, Solar Phys., 1, 242.

Swarup, G., et. al.: 1963, Astrophys. J., 137, 1251.

Swarup, G.: 1963, AAS - NASA Symposium on the Physics of Solar Flares, 179.

Tanaka, H. and Steinberg, J.L.: 1964, Ann. Astrophys., 27, 29.

Tsuchiya, A., and Nagane, K.: 1967, Solar Phys., 1, 121.

Underwood, D.: 1967, private communications.

Widing, K.G.: 1966, Astrophys. J., 145, 380.

Zheleznyakov, V.V.: 1962, Soviet Astron. - A.J., 6, 3.

Zheleznyakov, V.V.: 1964, Soviet Astron. - A.J., 7, 630.

Zlotnik, E. Ya.: 1968, Soviet Astron. - A.J., 12, 245.

Zlotnik, E. Ya.: 1968a, Soviet Astron. - A.J., 12, 464.

A SURVEY OF SOLAR ECLIPSE MEASUREMENTS IN THE 1-30 MM RANGE

D.L.Croom

S.R.C.Radio and Space Research Station, Slough

Bucks., U.K.

ABSTRACT

Solar observations in the 1-30 mm band have been made at a number of eclipses since 1945, and this paper summarises the main results of these measurements. The importance of these wavelengths to solar astronomy lies in the fact that radiation in this band originates at levels only a few thousand km above the visible surface of the sun, and hence gives information on the lowest layers of the solar atmosphere. This region (the chromosphere) is particularly important as it is the region in which solar flares originate. At these very short radio wavelengths it is possible to obtain a total eclipse, a phenomena which does not occur at metre and decimetre wavelengths.

1. INTRODUCTION

In comparison with the extensive solar studies that have been carried out at metre and centimetre wavelengths, the mm band (Fig. 1) or more specifically the wavelength region from 30 mm down to 1 mm (10-300 GHz), has until recent years been very largely ignored. Such work as was carried out prior to, say, 1965 consisted mainly of a few short periods of monitoring the sun for solar bursts (Hagen and Hepburn 1952, Coates 1966), together with a few eclipse studies, which will be described later. This situation has changed somewhat over the last few years, with both solar mapping and solar-burst monitoring being extended to millimetre wavelengths. This is a result of a combination of the facts that large-scale routine studies at the longer wavelengths are now well established, that the last few years have seen big improvements in

177

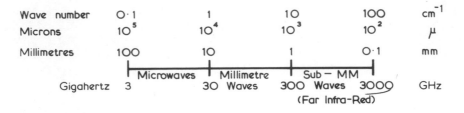

Figure 1   Relation of the millimetre band to microwave and infra-
            red wavelengths

mm techniques and in the commercial availability of components,
particularly reasonably good crystals, and finally, in an increasing awareness of the importance of these very short wavelengths to
solar physics.

     This importance arises because these waves are generated in,
and can propagate from, those regions of the sun where solar flares
originate, a few thousand km above the visible surface of the sun.
Although longer wavelengths may be generated in these regions (the
CHROMOSPHERE), the plasma density is such that they cannot propagate, and hence cannot be used for studying this part of the solar
atmosphere.   Millimetre wavelengths can be used for studying both
the pre-burst changes in emission from potential flare regions (the
slowly-varying component, Kislyakov and Salomonovich 1963, Tsuchiya
and Takahashi 1967), and also the very early stages of the flare
itself, the burst component (Croom and Powell 1969), and hence they
are of particular importance to the current intense international
interest in proton flares.   The basic solar emission (the quiet
sun component) is itself of some interest at these wavelengths, it
having recently been suggested that the effective temperature of
the sun has a minimum at $\lambda = 6$ mm (Fig. 2) due to small-scale
chromospheric turbulence (Kaplan and Tsytovich 1967), though recent
measurements by Gautier (1969) suggest that this minimum does not
exist.

2.   A SURVEY OF PRE-1966 ECLIPSE RESULTS

     Such interest as there has been in solar studies at these
wavelengths does, in fact, extend back to the very early days of
radio-astronomy, the first solar radio eclipse data being recorded
at a wavelength of 12.5 mm during the partial eclipse of 9 July
1945 (Dicke and Beringer 1946).   The results of this work showed
that at this wavelength the size of the sun's disc was not greatly
different from that at optical wavelengths.

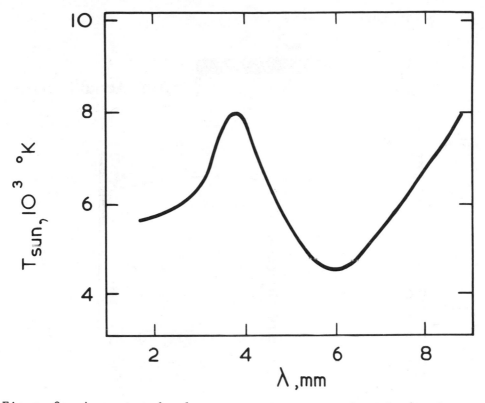

Figure 2   A suggested solar spectrum at mm wavelengths (Kaplan and
            Tsytovich 1967)

        At subsequent eclipses on 25 February 1952 and on 30 June 1954,
observations were carried out at $\lambda = 8.5$ mm (Hagen 1956) using a
fan-shaped beam 1° long by 8' wide, so that only a strip across the
centre of the solar disc was being observed (Fig. 3).   The resulting radial distribution curve derived from the measurements showed
a narrow bright ring around the disc, together with a small enhancement of emission towards the centre.   Hagen interpreted this result
in terms of a grass-like structure.   By a trial and error method he
found that a series of cool, dense pyramidal spicules, 3" wide at
the base, and on average 10,000 km in height, extending into a hot,
low-density gas, would give the observed brightness distribution.
On this interpretation the central rays come partly from the cool
spicules and partly from the hot inter-spicular gas, and hence will
record an equivalent temperature equal to the mean of the two.
Rays from nearer the limb come from the surfaces of the cooler, but
more highly-absorbing spicules, and at the limbs themselves the ray

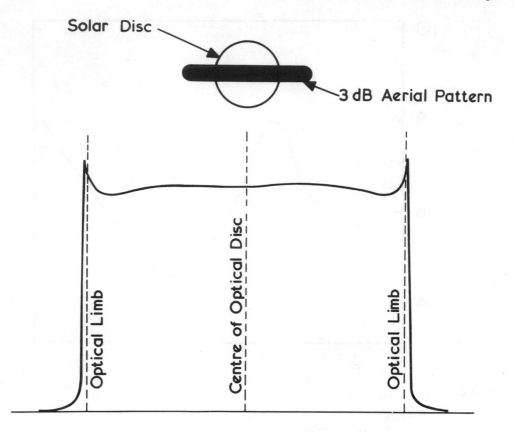

Figure 3   8.5 mm radial brightness of the sun deduced from the
            30 June 1958 eclipse (Hagen 1956)

path becomes tangential, absorption occurs in higher hotter
regions and hence limb-brightening occurs.

These results were supported by observations carried out at
millimetre wavelengths (unspecified) in the Ukraine during the
1954 eclipse, (Salomonovich, Pariiskii, Kangil'din, 1958).

Measurements at the eclipse of 19 April 1958 produced
negative results at 20 and 8 mm, though active regions were
detected at longer wavelengths (Molchanov et al. 1959).

The February 15 1961 eclipse was investigated by Tolbert and
Straiton (1962) at 4.3 mm wavelength.   The resulting occultation
curve showed some very interesting changes in slope which suggested

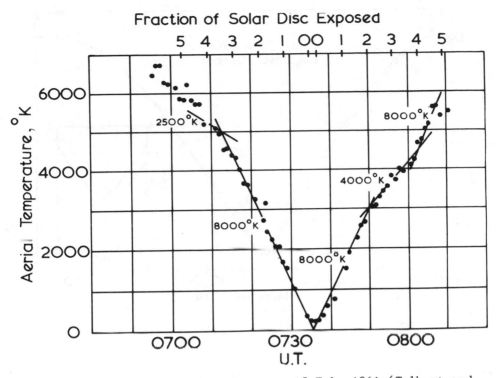

Figure 4   4.3 mm solar eclipse curve, 15 Feb. 1961 (Tolbert and
            Straiton 1962)

that extensive regions of the sun were at relatively very low
temperatures down to 2500°K (Fig. 4).

The same group carried out observations at 3.2 mm during the
20 July 1963 eclipse (Tolbert, Krause and Straiton 1964).
Although bad weather interfered with the results, there was a
suggestion of enhanced emission from prominences on the south-east
and south-west limbs.   No systematic limb-brightening was
observed, however.

3.   THE 20 MAY 1966 ECLIPSE AT MILLIMETRE WAVELENGTHS

The 20 May 1966 eclipse was observed in Greece as an annular
eclipse at 16 mm (Croom and Powell 1967), in Russia as a partial
eclipse at 16 mm (Apushkinskii et al. 1968), and in London as a
partial eclipse at 1.2 mm (Newstead 1969).

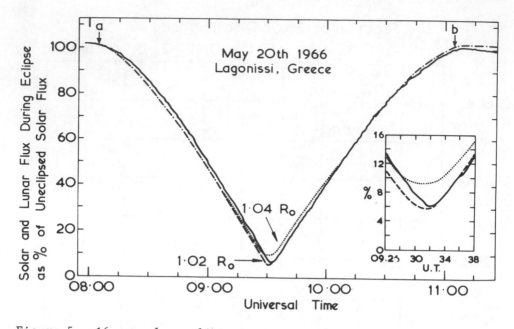

Figure 5   16 mm solar eclipse curve, 20 May 1966 (Croom and
              Powell 1967)

The 16 mm observations confirmed the existence of the weak
active region observed near the east limb at longer wavelengths
(Straka 1969), the estimated 16 mm flux being $8.3 \pm 1.5 \times 10^{-22}$ w
m^{-2} Hz, compared with 7.5 at 3.2 cm and 6.0 at 6 cm.   Other active
regions observed at the longer wavelengths were not detected at
16 mm.   The eclipse curve recorded by Croom and Powell is shown in
Fig. 5, together with some computed curves.   Figure 6 is based on
the Fraunhofer Institute solar map for the eclipse day.   The comparison of the measured and computed curves, together with the
differences between the times of optical and radio contacts,
suggests a maximum value of 10,000 km for the effective height of
emission of the quiet sun 16 mm radiation.

The 1.2 mm observations, which were made by scanning across
the lunar-solar interface with a 6'-7' aerial beamwidth showed the
existence of limb-brightening at this wavelength (Newstead, 1969).
Another interesting feature reported by Newstead was the occurrence
of a burst (Fig. 7) during the eclipse.   Newstead does not correlate this with any other solar phenomena, nor have any other radio
observers reported bursts during the eclipse.   However, according
to the Solar-Geophysical Data Bulletin issued in September 1966 by
the U.S. Environmental Science Services Administration, two optical

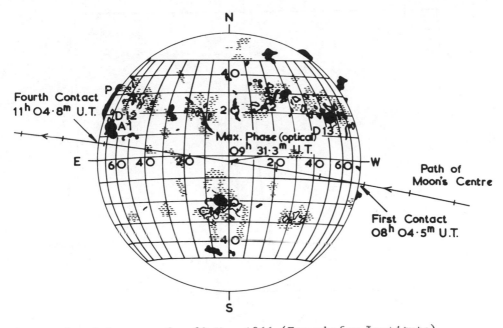

Figure 6   Solar map for 20 May 1966 (Fraunhofer Institute)

observatories (Arcetri and Bucarest) recorded a 1N flare at about
this time.   The relevant times as reported by Bucarest and the
approximate times derived from Newstead's reconstruction of the
burst are as follows:

	Optical	Millimetre
Start	1031	1030
Maximum	1048	1100
End	1204	1200

Newstead estimates the peak flux increase to be about 1.4% of the
total quiet sun solar flux, which if one assumes an effective disc
temperature of 6000°K, corresponds to a flux increase of approximately 1100 x 10^{-22} w m^{-2} Hz^{-1}.

ACKNOWLEDGEMENT

    This survey was prepared at the Radio and Space Research
Station of the Science Research Council and is published with the
permission of the Director.

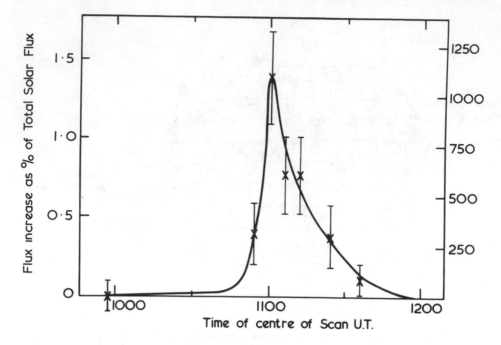

Figure 7   1.2 mm solar burst observed during the 20 May 1966
            eclipse (Newstead 1969)

REFERENCES

Apushkinskii, G. P.,   1968   "Radio-astronomical obser-
  Grebin, A. S., Enikeev, R. I.,   vations during the solar
  Levickenko, M. T. and   eclipse of 20 May 1966 at
  Nagnibeda, V. G.   wavelengths 0.8, 1.6, 3.3
                                         and 10 cm", Vesinik
                                         Leningradskogo Univ. Ser.
                                         Mat. Mekh. Astron. (USSR),
                                         No. 2, pp. 141-144

Coates, R. J.   1966   "Solar observations at milli-
                                         meter wavelengths", Proc.
                                         IEEE, 54, pp. 471-477

Croom, D. L. and   1967   "Solar radiation at 19.0 Gc/s
  Powell, R. J.   during the eclipse of May 20
                                         1966", Nat. 215, pp. 260-
                                         261

Croom, D. L. and   1969   "Outstanding solar radio
   Powell, R. J.   burst at 4.2 mm", Nat. 221,
                                  pp. 945-947

Dicke, R. H. and   1946   "Microwave radiation from the
   Beringer, R.   sun and moon", Ap. J., 103,
                                  pp. 375-376

Gautier, D.   1969   D.Sc. Thesis, University of
                                  Besancon, France

Hagen, J. P.   1956   "Radial brightness distri-
                                  bution of the sun at 8 mm",
                                  Solar eclipses and the
                                  ionosphere (Special Suppl.
                                  Vol. 6 to J. Atmosph. Terr.
                                  Phys.), Pergamon Press,
                                  London, pp. 253-257

Hagen, J. P. and   1952   "Solar outbursts at 8.5 mm
   Hepburn, N.   wavelength", Nature, 170,
                                  pp. 244-245

Kaplan, S. A. and   1968   "An interpretation of the
   Tsytovich, V. N.   dip in the spectrum of
                                  millimeter solar radio
                                  emission", Soviet Astron.,
                                  AJ-11, pp. 834-836

Kislyakov, A. G. and   1963   "Radio emission from solar
   Salomonovich, A. E.   active regions in the
                                  millimeter region", Soviet
                                  Astron. AJ-7, pp. 177-181
                                  (Trans. from Astron. Zhurnal
                                  40, pp. 229-234, 1963)

Molchanov, A. P. et al.   1959   "Preliminary results of radio-
                                  astronomical observations of
                                  annular solar eclipse, April
                                  19, 1958", Paris Symp. on
                                  Radio Astronomy (ed.
                                  R. N. Bracewell), pp. 174-
                                  175

Newstead, R. A.   1969   "Solar limb brightening and
                                  enhancement measurements at
                                  1.2 mm", Solar Phys., 6,
                                  pp. 56-66

Salomonovich, A. E., Pariiskii, Iu. N. and Khangil'din, U. V.	1958	"Observations of the total solar eclipse of June 30, 1954 at millimeter wavelengths", Soviet Astron. AJ-2, pp. 612-614 (Russian Vol. 35)
Straka, R.	1969	"Microwave spectral observations of coronal condensations", Proceedings of Nato Symposium on Solar Eclipses and the Ionosphere, Athens
Tolbert, C. W. and Straiton, A. W.	1962	"Observations of 4.3 mm radiation during the solar eclipse of February 15, 1961", Ap. J., 135, pp. 822-826
Tolbert, C. W., Krause, L. C. and Straiton, A. W.	1964	"Solar radiation at 3.2 mm during the July 20, 1963 eclipse", Ap. J., 140, pp. 306-312
Tsuchiya, A. and Takahashi, K.	1967	"Spectrum of slowly varying component of solar radio emission on millimeter wavelengths", Solar Phys., 2, pp. 104-106

DISTRIBUTION OF X-RAY EMISSION FROM THE SUN DEDUCED

FROM MEASUREMENTS OF IONOSPHERIC ABSORPTION

Demetrius   Boviatsos

University of Athens

During the annular solar eclipse of May 20,1966,over parts
of northern Africa and southern Europe,solar flux and ionospheric
absorption measurements were made from several points along the
central line of the path of the eclipse. Solar flux was measured
at various frequencies in the cm and mm bands.All the data obtained during the eclipse showed characteristic variations in flux
related to local active sources on the solar disk.Fig. 1 shows
the distribution of local sources on the solar disk measured in

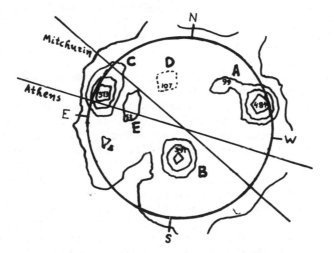

*Fig.1.Distribution of local active sources on the solar disk
and the path of the Moon following Athens and Mitchurin
eclipse geometry.*

Athens by Castelli on 2,695 MHz (private communication).Three sources giving an increase in flux were identified during this eclipse:
one on the west limb, the second in the middle of the disk and the
third on the east limb. Two sources of less importance were also
identified on the north-east part of the disk.

        Fig.2 shows the change of 10 cm flux per min,dF/dt,versus UT
obtained by the Greek and the Air Force Cambridge Research Laborato-
ries(AFCRL) group in Athens. Only important changes were considered,
changes of minor importance being neglected. The times of covering
by the Moon of the three local sources, corresponding to increased
flux,are also indicated.We can see that,following the Athens eclipse
geometry,the covering of local active sources coincides with changes in dF/dt.This is also observed as the sources are uncovered.

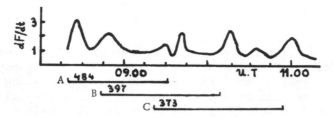

Fif.2.Changes in flux per minute versus time, and the times
       of covering of local active sources transmitting
       increased flux.

        Absorption measurements of the A1 type were made during the
eclipse with the Scaramanga ionosonde, near Athens. Fig.3a shows the
change of absorption per minute dA/dt,versus time,computed from the
absorption eclipse curve on 1.98 MHz.Because the radio telescopes
used by the AFCRL staff and the Greek group were located only a few
km from the ionosonde, the times of covering of the active centres
are the same as before. In fig.3a the covering of active sources
coincides quite closely with the maxima of the time-derivatives of
ionospheric absorption, and the same coincidence is also observed at
the times of uncovering.The time derivatives of flux,dF/dt,from the
active sources and of ionospheric absorption,dA/dt, are therefore
closely correlated. This indicates that the three main centers of
activity, denoted by A,B and C,are also centers of strong X-ray emission,producing absorption in the D-layer,as measured on 1.98 MHz.Di-
rect measurements on X-ray radiation made during the same eclipse by
Landini,Russo and Tagliaferri,using data from the US Naval Research
Laboratory 1965A satellite,show a large change in the band 1-8 Å at
the covering of the active source in the middle of the solar disk,
and also at the uncovering of the active source on the east limb.
These results support our observations,and we may assume that parts
of local sources emitting hard X-rays are causing the changes in the

D-layer absorption.

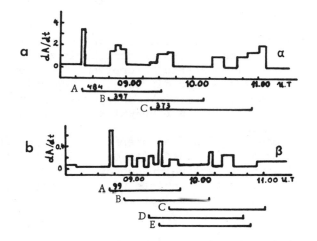

Fig.3a.dA/dt versus time-measured on 1.98 MHz,and the times
       of covering of the local active,sources transmitting
       hard X-rays (Athens).
Fig.3b.dA/dt versus time,measured on 3.85 MHz,and times of
       covering of the local active sources transmitting
       soft X-rays (Mitchurin).

     During this eclipse,absorption measurements of type A1 were
also made by Bischoff and Taubenheim at Mitchurin,in southern Bulga-
ria,which was also in the centre line of the eclipse path on the
ground.The eclipse geometry at Mitchurin is different from the geometry at Athens.The frequency of 3.85 MHz used by Bischoff and Taubenheim was almost double the frequency used in Athens;for this
reason the absorption curve obtained at Mitchurin refers to absorption conditions in the E-layer.Fig.3b shows the change of absorption
per minute dA/dt versus time,computed from the Mitchurin absorption
curve on 3.85 MHz.The dA/dt values are much less than those measured
at Athens.In this figure, the times of the three local active sources are indicated,following the eclipse geometry appropriate to Mi-
tchurin.Correlation between changes in absorption and the times of
the local active sources previously considered in Athens is poor.
In addition,there are changes in dA/dt not related to the three main
active sources.By examining the dA/dt maxima and the covering of the
solar disk following the Mitchurin geometry in more detail,we can
see that sources,A,B and C,considered as producing absorption in the
D-layer following the Athens eclipse geometry,did not produce any
effect in the E-layer.Only the parts of the corona and of the two
sources in the limb which are emitting soft X-rays are coincident
with the dA/dt maxima observed at Mitchurin.In addition,two other
centers of soft X-ray radiation,located in the north-east part of
the solar disk and denoted by D and E produced changes in dA/dt.

We may assume that sources emitting hard X-rays are responsible
for the absorption effect in the D-layer, and that soft X-rays, especially those located in the limb of the solar disk, are associated with absorption in the E-layer.

From these results it seems possible, during an eclipse, to
investigate the distribution of hard or soft X-ray sources on the
solar disk, by measuring ionospheric absorption at different frequencies selected to correspond to different ionospheric layers.
In general, the ionosphere, seems to be an adequate mean to detect
the X-rays radiation of the Sun, during extra solar events.

Solar flares produce sudden ionospheric disturbances (SID) by
emitting X-rays. Ionosondes provided with devices measuring almost
continuously the minimum reflective frequency (fmin.) of the ionosphere may be used in order to calculate with accuracy the time
variation $S(t)$ of the ionizing X-radiation, associated with the
solar flare.

Fig. 4 shows a continuous recording of fmin. produced by the
solar flare of 19.5.67 and obtained by the ionosonde of the Ionospheric Institute of the National Observatory of Athens. In order
to calculate $S(t)$ we start by ploting fmin(t) from the above continuous ionograme. Electron densities of the D-layer are proportional to the second power of the sum (fmin.+f_L) where f_L, the
gyrofrequency. From fmin(t) curve the ratio N/No electron density
during CID is computed, expressing the electron increase
due to the ionizing X-rays component of the solar flare. By assuming

Fig.4 Continuous recording of fmin produced by the solar flare
      of 19.5.67

$N/No = M(t)$ and $\tau_0 = 1/2\alpha.N_o$, where α is the recombination coefficient considered as constant throughout the SID, we found the relation :

$$S(t) = M^2 + 2\tau_0 \, dM/dt$$

from which $S(t)$ is related not only to $M(t)$ but also to the time τ_0. Time of reconstitution of the ionosphere τ_0 however, is dependant from several parameters seasonals and others. For this reason in assuming that after the flare the electron production rate $q \to q_0$ (q_0 = electron production rate before flare) and consequently $S(t) \to I$, we simplify the above relation as follows

$$2\tau_0 = I - M^2/dM/dt$$

in order to calculate τ_0

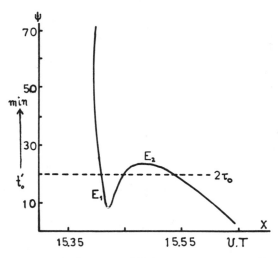

Fig.5. By plotting M(t) curve different values of τ_0 are determined The value of $t'_0 = 2\tau_0$ corresponds to the condition $E_1 = E_2$, equality of surface airs.

Fig. 6a shows $S(t)$ curve for the solar flare of May 19,1967 plotted following the above method from the continuous ionograms.

Previously   no direct test of the validity of the above method was possible. Only in the recent time when direct data on X-rays emitted from solar flares were obtained by the Explorer 33 satellite, this comparison was made possible. Fig. 6b shows $F(t)$ curve for the same solar flare of May 19,1967 plotted from direct data transmitted by Explorer 33. Absolute value of X-rays flux F in the 2-12 Å range expressed in $erg.cm^{-2}.sec^{-1}$ is given by the relation :

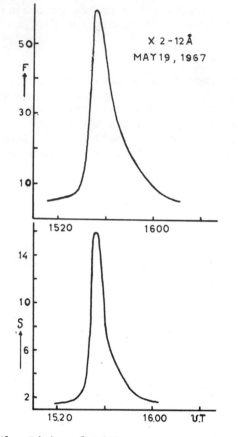

Fig. 6a.,6b. S(t) and F(t) curves for the flare of
19.5.67.

$$F = 1,8.10^{-6}.f(\delta).R$$

where R = corrected value of pulses by sec., δ = angle between rotation axis of satellite and satellite - Sun axis, and f(δ) coefficient of geometrical oblique incidance.

It is clear from the curves shape that a close correlation
exist between S(t) curve computed from the continuous ionograms
and F(t) plotted after direct value of X-rays flux in the 2-12 Å
transmitted by Explorer 33. Correlation coefficient between the
above curves is equal to 0.976.

A number of data on X-rays from solar flare, obtained by Explorer -33 were kindly communicated to us by Professor J.A. Allen
and Dr Krimigis of the Iowa University.

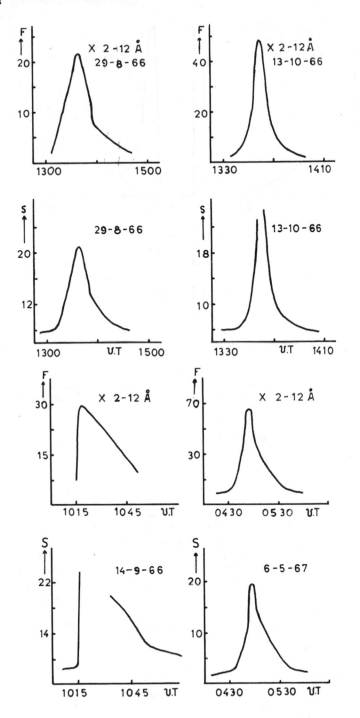

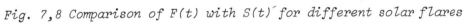

Fig. 7,8 Comparison of F(t) with S(t)´ for different solar flares

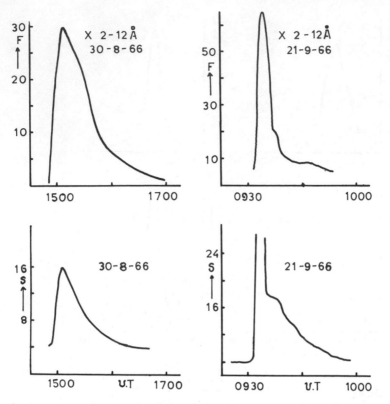

Fig.9.Comparison of F(t) with S(t) for solar flares of 30.8.66 and 21.9.66

    Fig.7,8 and 9 show a number of curves of S(t) and F(t)between
which the correlation remains as high as previously.It is than
evident that continuous recordings are an excellent mean to detect
the X-rays component of solar flares.This is a great advantage,
because continuous ionograms are easily available and also because
now when the method by continuous ionograms is well established
it would be possible to investigate a large number of solar flares
which took place in the past,when satellite did not existed and
only continuous recording by ionosondes of the type used by the
Ionospheric Institute of the National Observatory of Athens are
available.

REFERENCES

LANDINI,M.,RUSSO,D.,   *Nature, 211, 394 (1966).*
  and TAGLIAFERRI G.L.
BISHOFF,K., and   *J.Atmos.Terr.Phys. 29, 1063 (1967)*
  TAUBENHEIM J.
ELWER,G.   *Geophys.Res. 66, 391 (1961)*
FRIEDMAN,H.   *Space Res.,2,1021 (1961)*
BOVIATSOS,D.   *Thesis,Univ.of Athens, (1969)*
UNDERWOOD J., and   *Sol.Phys. 1, (1967)*
  MUNEY W.

ECLIPSE STUDIES BY ROCKET, SATELLITE, AND PARTIAL REFLECTION TECHNIQUES

D-REGION ELECTRON DENSITY MEASUREMENTS

DURING THE SOLAR ECLIPSE OF MAY 20,1966

J.A.Kane

Laboratory for Space Sciences,NASA Goddard Space

Flight Center,Greenbelt,Maryland,U.S.A.

ABSTRACT---Rocket (Underwood and Muney, 1967) and satellite observations of 2-8Å X-ray emissions indicate that the solar eclipse of May 20, 1966 occurred during a period in which the Sun was moderately active. The Goddard Space Flight Center conducted a series of small sounding rocket experiments to measure the ionospheric D-region electron density profile during successive stages of the eclipse. Compared to the uneclipsed profile, an order of magnitude reduction is observed in the D-region profile obtained near totality. From the electron density measurements together with the X-ray spectral data of Underwood and Muney it is concluded that on May 20, 1966, 2-8 Å X-rays were not the dominant source of D-region ionization below approximately 80 km.

INTRODUCTION

It is generally believed that during quiet Sun conditions the D-region of the ionosphere (i.e. the altitude interval between 70 and 85 km) is produced by the Lyman alpha ionization of nitric oxide. However, during periods of increased solar activity the flux of kilovolt X-rays penetrating the D-region increases by orders of magnitude while the Lyman alpha flux remains essentially constant. At some level of activity it can be expected that the production of D-region ionization by X-rays will exceed the Lyman alpha-nitric oxide contribution. At any given time however it is difficult to evaluate the relative

weights of these two ionization sources because the
nitric oxide concentration is a gross uncertainty.   An
alternative to this impasse might be provided by the
conditions at a solar eclipse.

The May 20, 1966 solar eclipse occurred during a
period in which the Sun was moderately active.   X-ray
photographs of the Sun obtained at White Sands, New
Mexico (Underwood and Muney, 1967) on the day of the
eclipse show three active regions for emissions in the
3-11 Å wavelength interval.   By the covering and uncovering of these spots the solar eclipse provides an ideal
situation for studying the effects of these X-rays on
the D-region ionization content.

The path of the May 20, 1966 solar eclipse as it
passed over Greece is shown in Fig. 1.   In collaboration
with the Ionospheric Institute of the University of
Athens, the Goddard Space Flight Center conducted a series
of ARCAS rocket experiments to measure the D-region electron density profile during successive stages of the
eclipse.   The experimental site was located on the beach
at Koroni in the southern Peloponese, while the rocket
launchings took place from a ship located approximately
3 km off shore.

The rocket firing schedule together with the pertinent parameters of the eclipse are given in Table 1.   The
eclipse data refers to a location close to the peak of
the rocket trajectory.

EXPERIMENTAL METHOD

The electron density profiles were determined from
radio propagation measurements.   The well known Faraday
rotation and differential absorption effects were measured on cw radio transmissions at 2·73 and 4·03 Mc/s.
For each frequency a linearly polarized signal was radiated from a ground based antenna and received on a separate linearly polarized antenna in the rocket.   The
mechanical spin of the rocket of about 25 c/s was used
to rotate the receiving antenna through the polarization
pattern of the arriving wave.   The telemetered signal
strength exhibited a fading pattern which was the sum of
the rocket spin frequency and the ionospheric Faraday
rotation.   A comparison of this fading pattern with the
mechanical spin period yielded the altitude variation of
the plane of polarization.   The mechanical spin period
was obtained from the fading pattern of the 240 Mc/s

FIG. 1. PATH OF THE MAY 20, 1966 ECLIPSE IN GREECE
Rocket experiments were conducted at Koroni.

TABLE 1

| | | Local time | | | Illumination | |
		(UT + 2:00)	Zenith angle	Lyman alpha*	X-ray**	Solar index***
May 20	First contact	10:00				114
	Zorba II	10:00	37·0°	1·0	1·00	
	Zorba III	10:45	28·0°	0·53	0·66	
	Eclipse maximum	11:25				
	Zorba IV	11:30	21·5°	0·09	0·10	
	Zorba V	12:15	17·2°	0·66	0·83	
	Zorba VI	13:00	18·0°	1·00	1·00	
	Last contact	13:00				
May 21	Zorba VII	11:30	21·5°			121

*Proportional to uneclipsed solar disc area.
**2-8Å X-ray band.
***2800 Mc/s flux (High Altitude Observatory, Boulder, Colorado).

telemetry carrier frequency as observed on the ground with a linearly polarized receiving antenna.

Differential absorption data is also obtainable from the fading pattern of the received signal strength. This follows from the fact that the maxima in the signal strength represent the sum, while nulls represent the difference in the ordinary and extraordinary components of the linearly polarized wave.

Under the condition of quasi-longitudinal propagation, the plane of linear polarization, defined by the angle φ, rotates with rocket altitude z according to an expression of the form:

$$\frac{d\varphi}{dz} = N_e(z)F(\omega,\omega_H,\nu(z))$$ (1)

where $N_e(z)$ is the electron density and F is a calculated function of the exploring frequency ω, the electron gyro frequency ω_H and the collision frequency ν. The explicit form of F involves the Dingle integrals of the generalized Appleton-Hartree formula (see for example, Sen and Wyller, 1960).

Differential absorption can be related to the ionospheric parameters by a similar expression. Denoting the received signal strengths of the two polarization modes as E_0 and E_x the altitude variation of the logarithmic ratio $\ln(E_0/E_x)$ can be expressed as

$$\frac{d}{dz}\ln(E_0/E_x) = N_e(z)G(\omega,\omega_H,\nu(z))$$ (2)

where again $N_e(z)$ is the electron density and $G(z)$ is a calculated altitude dependent function involving the Dingle integrals. Before either $F(z)$ or $G(z)$ can be calculated it is necessary to assume a collision frequency model. In Table 2 is listed the collision frequency model used in the present work. This model was based on the mean monthly pressure data of Kantor and Cole (1965) and the relationship of Phelps (1960)

$$\nu = 6 \cdot 28 \times 10^7 p \ (sec^{-1})$$ (3)

where p is the atmospheric pressure in millibars.

TABLE 2

Z	ν	N_{II}	N_{III}	N_{IV}	N_V	N_{VI}	N_{VII}
67	$5 \cdot 27 \times 10^6$				40±50	130±50	120±170
69	$3 \cdot 89$	145±60			60±50	220±50	230±50
71	$2 \cdot 83$	240±60	160±90		160±50	330±50	380±50
73	$2 \cdot 05$	345±50	350±165	15±60	350±70	550±85	560±60
75	$1 \cdot 47$	430±100	480±50	28±50	450±50	620±60	690±90
77	$1 \cdot 04$	500±260	520±50	48±50	550±50	690±50	780±100
79	$7 \cdot 22 \times 10^5$		550±50	77±50	660±50	740±80	850±50
81	$5 \cdot 15$		580±50	115±50	780±50	1000±80	960±180
83	$3 \cdot 61$		900±100	155±50	1100±80	2000±280	1150±180
85	$2 \cdot 51$		1800±350	225±50	2000±370	2900±150	2000*
86							3400*
87	$1 \cdot 75$		3500±1700	660±80	2550±880	4500±325	7000*
88·5						10,000*	
89	$1 \cdot 22$		6000±670	1500±280	4600*		
90					6600*		
91	$8 \cdot 49 \times 10^4$		11,000±3750	3300±380	10,000*		
92				5700±300			
93	$6 \cdot 03$		22,000±13750				

*Based on $4 \cdot 0$ Mc/s Faraday data only

ELECTRON DENSITY RESULTS

Electron density profiles were obtained for all six rocket flights. The experimental method yielded four sets of data:

1. Faraday rotation on $2 \cdot 73$ Mc/s (240 Mc/s reference).
2. Faraday rotation on $4 \cdot 03$ Mc/s (240 Mc/s reference).
3. Difference Faraday ($2 \cdot 73$ Mc/s compared to $4 \cdot 03$ Mc/s).
4. Differential absorption on $2 \cdot 73$ Mc/s.

The relative quality of any one of these sets of data varied from flight to flight. This was due to both external and internal causes. Externally some radio interference on $2 \cdot 73$ Mc/s was encountered. Internally, the radiation pattern of the 240 Mc/s transmitting antenna apparently underwent some detuning during flight. This resulted in a Faraday reference signal too crude for the measurement of the small electron density in the lower altitude region. In this region the electron density profile was obtained from the difference Faraday (i.e. the $2 \cdot 73$ fading pattern compared with the $4 \cdot 03$ pattern) and differential absorption on $2 \cdot 73$ Mc/s.

Figure 2 shows the spread on the electron density values derived from three sets of data obtained on flight labeled Zorba VI. A similar spread for each Zorba flight determined the uncertainty assigned to the final electron density profiles given in Table 2. The Zorba trajectories were determined from an empirical relationship between peak time t_p and peak altitude z_p. t_p is accurately determined as the time at which the maximum Faraday rotation is observed. The relative accuracy on the Zorba trajectories is estimated to be $\pm 0 \cdot 5$ km, while the systematic error on the absolute values of z_p is estimated to be less than $\pm 2 \cdot 0$ km. The individual profiles of Table 2 are summarized in Fig. 3. This shows that the effect of the eclipse upon the D-region electron density is quite large near totality.

X-RAY INFORMATION

In Fig. 4 are shown maps of the eclipsed Sun as viewed from near the peak of each Zorba rocket trajectory. These maps (which were kindly provided by A.C. Aikin) show the location of the regions of 3-11 $\mathring{A}$ X-ray emissions

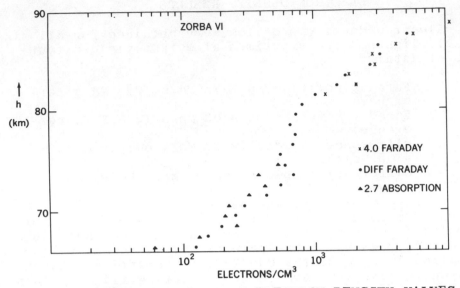

FIG. 2.   SPREAD ON ZORBA VI ELECTRON DENSITY VALUES
DEDUCED BY THREE SEPARATE METHODS

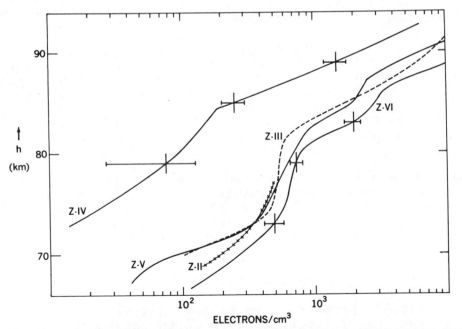

FIG. 3.   ELECTRON DENSITY PROFILES OBTAINED DURING THE
COURSE OF THE MAY 20, 1966 ECLIPSE

as determined by Underwood and Muney (loc. cit.). Also,
by a fortunate coincidence, the eclipse was observed in
the 1-8 Å X-ray band by a photometer aboard the SOLRAD
8 satellite. From this data Landini et al. (1966) obtained the relative weights of the West, Central and East
X-ray spots as 3,2 and 1 respectively. These weights
together with the maps of Fig. 4 determined for each
Zorba flight the relative X-ray illumination given in
Table 1.

The crucial X-ray information however, is the
spectral distribution within the 2-8 Å band. The spectrum which is given in Table 3 (J. H. Underwood, private
communication) was obtained at White Sands at 1510 UT
(i.e almost 4 hr after the end of the eclipse in Greece).
This spectrum together with the photoionization tables
of Ohshio, Maeda and Sakagami (1966) enable one to calculate $q_x(z)$, the altitude dependent ion production function due to X-rays alone. The results of these calculations for two zenith angles are given in Table 4.

TABLE 3

λ (Å)	2	4	6	8
ergs/cm² sec Å	$5 \cdot 0 \times 10^{-7}$	$7 \cdot 5 \times 10^{-6}$	$6 \cdot 5 \times 10^{-5}$	$5 \cdot 5 \times 10^{-4}$

TABLE 4

Z (km)	80	75	70
$q_x(\chi = 0°)$	$2 \cdot 9 \times 10^{-1}$	$4 \cdot 7 \times 10^{-2}$	$1 \cdot 0 \times 10^{-2}$
$q_x(\chi = 30°)$	$1 \cdot 9 \times 10^{-1}$	$3 \cdot 1 \times 10^{-2}$	$7 \cdot 0 \times 10^{-3}$

DISCUSSION

As indicated in Table 1, from one Zorba flight to the
next, the change in X-ray flux is almost identical to the
change in the Lyman alpha illumination. This situation
prevents us from seeing a pure X-ray effect in a comparison of the Zorba electron density profiles. However,
granted certain generally accepted assumptions, we can
show that the X-ray spectra presented in Table 3 could not
have dominated the production of ionization below approximately 80 km. The argument is as follows:

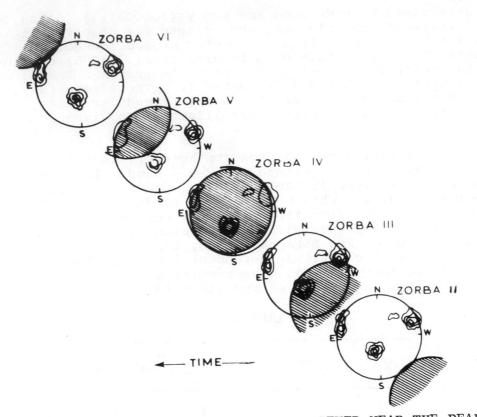

FIG. 4.   OBSCURATION OF THE SUN AS VIEWED NEAR THE PEAK
OF EACH ZORBA TRAJECTORY Included in each map are
isopleths of X-ray emission in the 3-11 Å band.

(See erratum, page 210.)

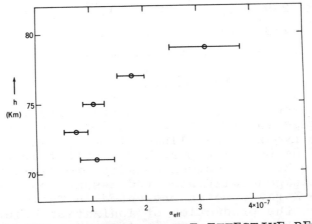

FIG.5. ALTITUDE VARIATION OF THE EFFECTIVE RECOMBINATION
COEFFICIENT, ASSUMING X-RAYS TO BE THE ONLY SOURCE OF
IONIZATION.

For the conditions existing at the end of the eclipse (i.e. $dN/dt = 0$) we can calculate an effective recombination coefficient $\alpha_{eff} = q_x/N^2$. We use the Zorba VI electron density profile and an ion production $q_x(\chi = 18°)$ obtained from the data given in Table 4. The results of this calculation are shown as a function of altitude in Fig. 5. Here the indicated uncertainty bars reflect the uncertainty in the electron density determinations only. Uncertainties in the Zorba VI trajectory or in the parameters entering into the calculation of q_x affect the magnitude of α_{eff} but do not significantly alter the α_{eff} altitude variation. Fortunately it is only this altitude dependence of α_{eff} which concerns us. Generally it is assumed that α_{eff} can be expressed as $\alpha_{eff} = (1 + \lambda)(\alpha_d + \lambda\alpha_i)$. Here $\lambda \equiv N-/N_e$ is the ratio of the densities of negative ions to electrons and α_d, α_i are the constant dissociative and mutual ion recombination coefficients respectively. Furthermore, in current D-region theories, λ is a monotonically decreasing function of altitude. It is therefore required that α_{eff} also be a decreasing or constant (depending on the magnitude of λ) function of altitude. Since the values of α_{eff} in Fig. 5 do not show a decreasing or constant altitude dependence within the experimental accuracy we conclude that the X-ray spectral values of Table 3 are insufficient to account for the ionization of the D-region below approximately 80 km on 20 May 1966. This conclusion is in disagreement with the conclusion deduced by Bowling, Norman and Wilmore (1967) from rocket-borne Langmuir probe measurements of the eclipse.

On 20 May the integrated energy flux in the 1-8 Å band was 5×10^{-4} ergs/cm² sec. On the following day at the time of the Zorba VII control shot the integrated energy flux in the 1-8 Å band had increased to 1.2×10^{-3} ergs/cm² sec (R.W. Kreplin, private communication). A comparison of electron density profiles for Zorba VII and Zorba VI listed in Table 2 shows that this increased X-ray flux apparently did not lead to any pronounced increase in the D-region electron density values.

CONCLUSION

On 20 May 1966 an ionizing radiation source other than 1-8 Å X-rays dominated the formation of the D-region below approximately 80 km. Implicit in this conclusion are the assumptions:

1. The X-ray spectra did not change significantly in the 4 hr between Zorba VI and the X-ray measurements at White Sands.

2.   An effective recombination coefficient can be written as $\alpha_{eff} = (1 + \lambda)(\alpha_d + \lambda\alpha_i)$ with α_d, α_i constant between 70 and 80 km.

3. λ is a monotonically decreasing function of altitude.

4.   The experimental values of electron densities and X-ray spectrum together with the calculated q_x values are correct.

Acknowledgements-This work was made possible through an international cooperative program between NASA and the Greek National Committee for Space Research.   We are especially grateful to Professor Michael Anastassiadis for valuable assistance with the field operations.   Mr. E. Bissell was the NASA Project Manager.

The communication of unpublished data by J. E. Underwood is gratefully acknowledged.

REFERENCES

BOWLING, T.S., K. NORMAN and A.P. WILMORE, D-region measurements during a solar eclipse. Planet. Space Sci. 15, 1035-1047 (1967).

KANTOR, A.J. and A.E. COLE, Monthly atmospheric structure surface to 80 km. J. appl. Met. 4, 228-247 (1965).

KREPLIN, R.W., private communication (1967).

LANDINI, M., D. RUSSO and G.L. TAGLIAFERRI, Solar eclipse of May 20, 1966 observed by the Solrad 8 Satellite in X-ray and ultra-violet bands. Nature, Lond. 211, 394 (1966).

OHSHIO, M., R. MAEDA and H. SAKAGAMI, Height distribution of local photoionization efficiency. J. Radio Res. Labs Japan 13, 70, 245-577 (1966).

PHELPS, A.V., Propagation constants for electromagnetic waves in weakly ionized air. J. appl. Phys. 31, 1723-1729(1960)

SEN, H.K. and A.A. WYLLER, On the generalization of the Appleton-Hartree magnetoinic formulas. J. Geophys. Res. 65, 3931-3950 (1960).

UNDERWOOD, J.H. and W.S. MUNEY, A glancing incidence solar telescope for the soft X-ray region. Solar Phys. 1, 129-144 (1967)

UNDERWOOD, J.H., private communication (1968)

ERRATUM - Figure 4 is incorrect. The correct figure may be obtained by rotating each of the five solar X-ray maps clockwise through 39.4°.

PARTIAL REFLECTION MEASUREMENTS ON THE D-REGION

DURING THE MAY 20,1966 SOLAR ECLIPSE

Emmanuel Tsagakis

University of Athens

During the solar annular eclipse of 20th May 1966,measurements of partial reflections from the D-region were made at a permanent out-station on Crete previously established by the Norwegian Defence Research Estbl.in collaboration with the Electronic Division of the Physics Department of the University of Athens.

The observation site was positioned about 200 km from the centre line of the eclipse,near Iraklion.At this station the eclipse lasted from 1002-1305 LMT with a maximum occultation of 95% at 1130 LMT.

Fig.1 shows the eclipse central path and also the location of two others sites in Koroni (South Peloponesse) and Scaramanga(Athens) from where measurements were performed during this same eclipse,by methods other than the partial reflection technique.

It is our purpose to compare experimental results on electron densities in the D-region,obtained during the eclipse,with two different techniques : a) Partial reflection and b) rockets, and also to make some assumption from absorption measurements of type A1,on the ionization produced by X-rays cources in the solar disc at different heights of the D-region.

Partial reflection measurements were performed in Crete by Haug Thrane of the Norwegian Defence Research Estbl. and myself of the University of Athens. Rockets measurements were performed by Kane of the NASA Goddard Space Flight center from Koroni in the South Pelloponese some 300 km NW from Iraklion, and A1 absorption measurements, by the University of Athens team in the Scaramanga

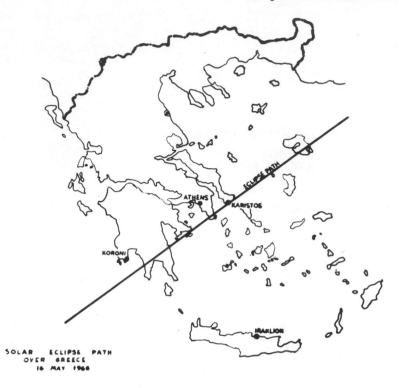

SOLAR   ECLIPSE   PATH
OVER   GREECE
16 MAY 1966

Fig. 1. Eclipse paths and location of measuring centers

ionosonder some 350 km North from Iraklion.

    It is then important to discuss first if it would be consis-
tant to compare results obtained from distant places and especially during the disturbed period of an eclipse. Under normal conditions the comparison of ionospheric parameters measured at locations spaced apart hundreds of kilometers is permissible. This
may also be true even during the disturbed conditions of an eclipse,
as is shown for instance when comparing the variations observed
during the above eclipse in Athens some distance of about 200 km
The possibility of comparing values especially in the D-region
depends mainly on the eclipse geometry and the location of measuring centers around the eclipse path. Measurements made during
the first quarter of the eclipse by the S-66 satellite which crossed the central eclipse path from South to North, showed that the
difference in electron content refered to the F region between
35°N (Iraklion) and 38°N (Athens) is reduced only by a factor of
0.95. It is then possible without a great error to attempt to
correlate electron densities measured by the partial reflection

technique in Iraklion with electron density values obtained from
direct rocket measurements in Koroni,and absorbtion values obtained by A1 pulse technique by the Scaramanga ionosonder in Athens.

Partial reflection experimental technique

The principle of the partial reflection method is briefly that a
linear polarized pulsed high frequency radio wave is transmitted
vertically and part of the energy is scattered back to earth from
irregularities in the D-region.

The receiver system separates the two magneto-ionic components
of the seattered signals and displays the amplitude of each component as a function of height on an oscilloscope screen.Photographs
of the screen are taken automatically at the rate of one per second.
By measuring the relative amplitude Ax/Ao of the extraordinary
and ordinary component as a function of height the electron density
profile can be deduced.

Our equipement in Crete proved it possible to deduce electron
densities at altitudes 70-90 km (Fig.2)

Fig.2.*General view of the partial reflection measuring
center in Iraklion Crete.*

Measurements of partial reflection were made by the Norwegian
group and ourselves during the control period 7-21 May 1966,on
every half hour from ground sunrise to ground sunset,each measurement consisted of about 500 amplitude height records,which are

sufficient to determine one electron density profile.During the
eclipse the camera was operated continuously producing sufficient
data to give the electron density profile for every 10 minutes.

In the analysis of the data the generalized magneto-ionic
theory has been used and it has been assumed that only irregularities
in electron densities are responsible for the partial reflection.The
collision frequency model used in the analysis is based upon the
assumption that collision frequency is proportional to pressure.
The model is given by Thrane and al.(1968).

Partial reflection results during the May 20,]966 solar eclipse

The analysis of partial reflection results was limited to the
altitude range 74-80 km where we feel the most reliable data comes
from Fig.3 and Fig.4 show the variation of the electron density
with the eclipse time at 74 km and 80 km.

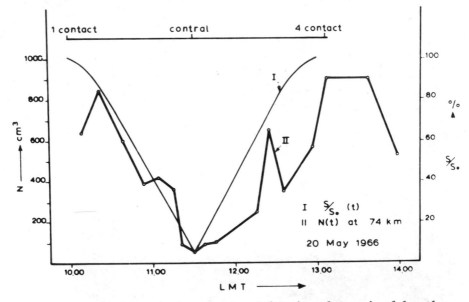

*Fig.3.Variation of the electron density determined by the
partial reflection technique versus eclipse time at
the height of 74 km.*

The scatter in the variations of electron density is of the
same order as the ecperimental uncertainty of the method. A smooth
variation through the eclipse will thus be consistant with the
observations within the experimental uncertainties.

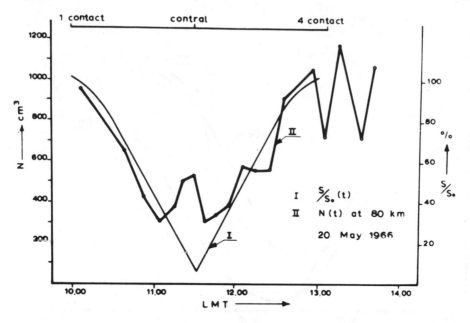

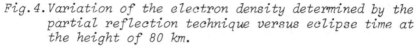

Fig.4.Variation of the electron density determined by the partial reflection technique versus eclipse time at the height of 80 km.

For all the above reasons we will avoid to try any correlation between the covering and uncovering of active centers present during the eclipse day in the solar disc with values obtained by partial reflection technique and we will consider any variations shown as purely accidental.

Fig.5 shows the smoothened electron densities values measured by the partial reflection technique at 81 km and 74 km versus time and in comparison with the optical occultation curve.At 74 km electron densities change by a factor of 17 throughout the eclipse.At 81 km the change is much less approximately a factor of 3. This difference in the lowest values of electron densities between these two heights of the D-region shows that for the altitude of 74 km the ionizing agent has a close dependance with the optical occultation of the solar disc and the active sources distributed on its surface while for 80 km an additional source of ionization than the solar disc alone, has to be considered,in order to justify higher values of N at the maximum of occultation of the solar disc. This statement is supported also by difference in the time delay between optical occultation and electron densities curves,indicating an effective recombination coefficient much higher in 74 km than in 81 km.

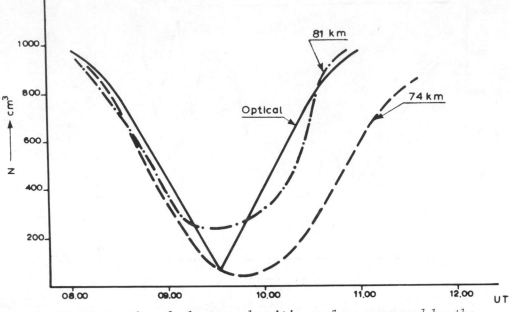

Fig.5.Smoothened electron densities values measured by the
      partial reflection technique at 81 and 74 km versus
      time and in comparison with the optical occultation
      curve.

Rocket measurements during the May 20,1966 solar eclipse

        Dr.Kane of the Goddard Space Flight center,proceeded during
the May 20, 1966 solar eclipse with a series of rocket launching
from Koroni in order to obtain altitude profiles up to 90 km of
the electron density as a function of the solar obscuration.From
his measurements some knowledge of the ionospheric time constant
and effective recombination coefficient were inferred and also ionization of the D-layer produced by hard X-rays was investigated.

        In the present we will discuss the results obtained by Dr.Kane
after the launching of Zorba I to VII compared with the results
on electron densities profiles determined by the partial reflection
technique.The distance between Koroni in the Southeast end of
Peloponnese and Iraklion may be considered as small and comparison
of results reliable.

Following the bibliography on the subject it seems that it is the
first time when such a direct comparison of results from the same
part of the ionosphere on electron content profiles was mede possible.
        Fig.6 shows after Dr.Kane the succesive launching of rockets

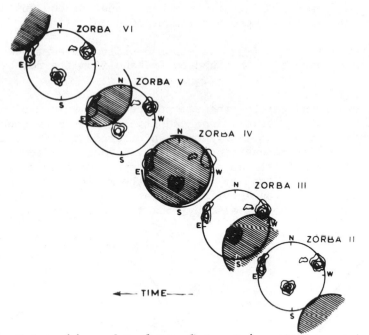

Fig.6.*Launching of Zorba rockets at important moment of occultation of the solar disc during the May 20,1966 solar eclipse.* (See erratum, page 210.)

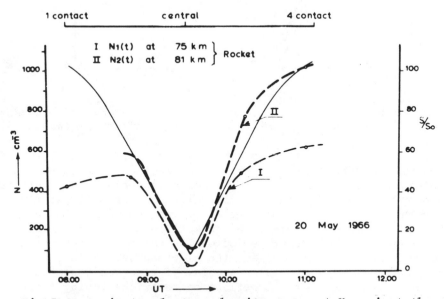

Fig.7.*Approximate electron density curve at Koroni at the heights 81 and 75 from the rockets launching.*

Zorba at the eclipse beginning at the maximum,and the end,and also
at two other intermediate important moment of occultation.

Fig.7 shows the approximate electron density curve at Koroni
at the heights 81 and 75 km obtained by Dr Kane from the rocket
launchings.

We observe again that for the same percentage maximum of occultation of the solar disc,electron densities measured by rocket
technique at 81 km are largely higher than electron densities at
75 km.For 75 km electron densities change by a factor of about
17 and for 81 km by a factor of about 6. Electron density values
however are higher at 81 than at 75 km for the maximum of occul-
tation.The hypothesis of an agent of ionization additional to the
solar disc alone is supported also by rocket measurements.

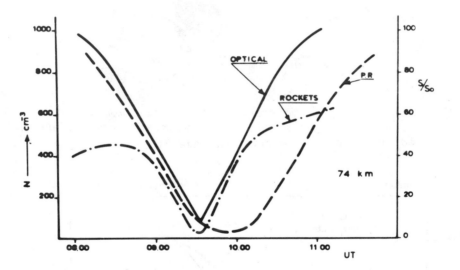

Fig.8.Electron densities eclipse curves at 74 km obtained
by both the rocket and partial reflection technique.

Fig.8 shows the electron densities eclipse curves at 75 km
obtained by both the rocket and partial reflection technique.At
maximum of optical occultation of the solar disc electron densities
values measured by both technique are quite similar.At least for
this percentage of occultation partial reflection is an equally
reliable technique as the direct by rocket.For other percentages
of the covering or uncovering period of the eclipse we observe no
time delay between the optical and the electron densities measured
by the rocket technique and a large delay of electron densities
values measured by the partial reflection technique during the second half of the eclipse.The physical behavior of the ionized layers
in the D-region is to present a timedelay versus the optical occul-

tation. It seems then that values obtained by the partial reflection
technique during the May 20,1966 solar eclipse,being more numerous,
have to be considered as more reliable.

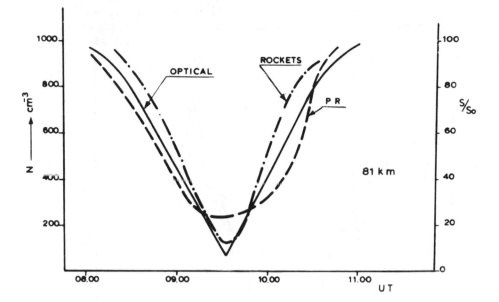

*Fig.9.Electron densities eclipse curves at 81 km obtained
by both the rocket and partial reflection technique.*

        Fig.9 shows the electron densities eclipse curves at 81 km.ob-
tained by both the rocket and partial reflection technique.At the
maximum of optical occultation electron densities measured by both
techniques are now higher than values obtained for the lower altitude of 74 km but no essential time delay is observed.In comparing
the situation in the ionized layers at 74 and 81 km we may accept
the statement made by Anastassiadis and Boviatsos following this
measurements that for the lowest level of 74 km ionization is produced by the penetrating radiation of hard x-rays emitted by the
three active sources distributed in the surface of the solar disc.
Their occultation by the Moon was successive and produced in the
D-layer a prompt effect.But the observed time delay during the second half of the eclipse is justified partly by the geometry of
their uncovering, and partly by the high value of the recombination
coefficient.For the 81 km altitude, the contribution from the above
active sources to the ionization is sharply reduced.The ionizing
agent is now Lyman - a and soft X-rays radiated not only by a
quasi uniform solar disc but also from the corona, as indicated by
Anastassiadis following his measurements in the F1 Layer.

Electron density profiles and absorption

The possibility to compare absorption measurements made in
Athens with electron density profiles measured by rockets in Koroni
and partial reflection technique in Iraklion, was discussed in a
previous paragraph.This depends mainly on the geometry of considered measuring centers, versus the eclipse path.In our case it
was proved that this was possible.

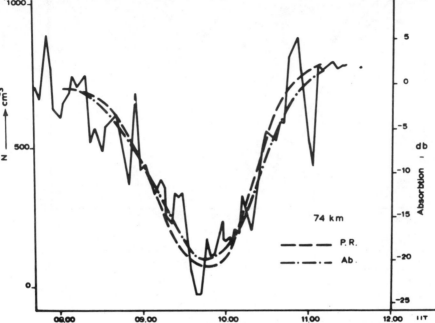

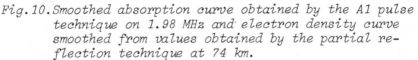

*Fig.10.Smoothed absorption curve obtained by the A1 pulse
technique on 1.98 MHz and electron density curve
smoothed from values obtained by the partial reflection technique at 74 km.*

Fig.10 shows plotted together the smoothed absorption curve
obtained by the A1 pulse technique by the ionosonde of Scaramanga
in Athens with the electron density curve smoothed from partial
reflection technique in Iraklion.All above electron density curves
are referred at a height of 74 km, where we feel we have the most
reliable data. By averaging in time, electron density values by
partial reflection technique, have been obtained every 10 minutes.
The absorption values are averaged every 3 minutes. This integration time is somewhat short for absorption measurements,but it is
very valuable for active centers distributed in the solar disc
investigations.

In order to compare the variations, electron densities and
absorption values are normalized to give the best fit.First, at the

start of the eclipse and second at the middle of the eclipse path,
The normalization implies that a change in the total absorption of
10 db corresponds to a change in electron density of about 600 elec.
cm. 3 both at 74 and 80 km. An estimate shows that about 25 per
cent of the absorption at 1.98 Mc/s takes place in a height range
between 70 and 80 km the rest is non deviative absorption above
80 km height and deviative absorption close to the reflection level. Little absorption seems to take place below 70 km.The height
of total reflection for 1.98 Mc/s signals varied from 92-96 km
throughout the eclipse,using informations on electron densities at
the height range from rocket experiments after Jespersen (1969).

A study of Fig.10 shows that with this normalization at 08.00
09.00 UT the general variation of the absorption is very similar
to the variation of electron density at 74 km measured by partial
reflcotion tecnique.This similarity did not exist for curves of
electron density compared with the absorption curve at 80 km as is
shown in Fig.11.

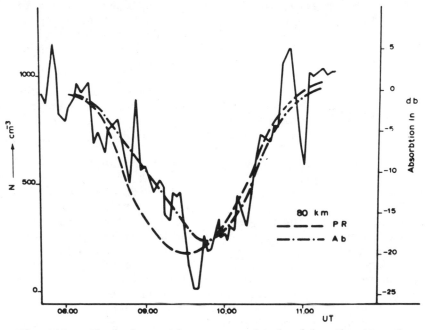

Fig.11Smoothed absorption curve obtained by the A1 pulse
technique on 1.98 MHz and electron density curve
smoothed from values obtained by the partial refle-
ction technique at 80 km.

From the above curves we can assume that the single frequency A1
absorption measurements give a good image of the variation of
electron density at a height level as low as 74 km.

A quick review of the absorption and the electron density curves shows that in complete contrast with the variations obser- ved in the electron density curves which for the greatest part must be considered as accidental, variations of the absorption curve are closely related with active sources distributed on the solar disc. Their covering and uncovering, produced very definite variations and moreover following a method of investigation indi- cated by Anastassiadis and Boviatsos it would be possible to indi- cate not only the influence of hard X-rays active centers but also their range of wavelengths.

Dr.Kane following his measurements during the May 20,1966 so- lar eclipse concluded that $1 - 8$ Å X-rays were not the dominant source of D-region ionization below 78 km.His statement is based on the calculation of a variable effective recombination coeffi- cient : $a_{eff} = q/N^2$ from rocket measurements on electron profile and the ion production function q due to the X-rays alone. Following Zorba results he found that the effective recombination coefficient in the D-region was not a decreasing function of alti- tude and since values of a_{eff} do not satisfy this altitude depen- dance, hard X-rays alone, do not produce the D-region ionization below 78 km the altitude at which the decreasing dependance chan- ges to increasing.

This conclusion is in disagreement with the conclusion dedu- ced from Bowling et al. from rocket born Langmuir probe measure- ments of the eclipse. Furthermore our own results are also in disagreement with the above statement of Dr.Kane.

The close correlation between the absorption curve on 1.98 Mc/s and partial reflection eclipse curve for 74 km and also the rather poor correlation between the above absorption curve and the partial reflection curve in 80 km indicate that it is a differen- ce in the ionizing agent for these two altitudes. The statement that the ionization of the D-region is not due to X-rays alone is correct. The only important difference consistant also with the measurements made by Anastassiadis and Boviatsos during the same eclipse is that active sources distributed in the solar disc sur- face are responsible for the ionization of layers below 78 km and that for higher altitudes, soft X-rays produced not only by the solar disc but also by the corona at the limbs are the do- minationizing agent.

REFERENCES

Thrane E.V.,Haug A.,Bjelland B.,   *(1967) J. Atmos.Terr.Phys.30,135*
  Anastassiades M.,Tsagakis E.
Anastassiades M.,Boviatsos D.,   *(1968) Nature,219,No 5159,1139*
Bowling T.S.,Norman K.,Willmore   *(1967) Planet Space Sci.15,1035*
Haug A.,Thrane E.V.   *Private Communication*
Anastassiades M.,Proceedings of   *(1966) Report No11A009*
  workshop on 20 May 1966,Solar
  eclipse
Kane J.A.,*Solar eclipse papers presented at conference on Meteo-*
  rological and Chemical factors in D-region Aeronomy-Urbana Il-
  linois Sept.,1968

ELECTRON CONTENT MEASUREMENTS BY BEACON S-66 SATELLITE

DURING THE MAY 20, 1966, SOLAR ECLIPSE

Demetrius Matsoukas

University of Athens, Electronics dp.

Introduction

During annular solar eclipse of May 20,1966, over the Northern part of Africa and the Southern part of Europe, it was possible for the first time, to investigate the effect of an eclipse on the to‐ tal electron content of the ionosphere, by using the satellite tech‐ nique. KLOBUCHAR, WHITNEY (1964),HAWARD et al.(1964) also observed the effect of the eclipse on the total electron content by using the moon reflection technique.

At 08.40 U.T. (the time of nearest approach over Athens),the S-66 beacon satellite crossed the solar eclipse area,following a path from North to South as shown in Fig.1.

The maximum obscuration in the F-region over Athens happened at 90.31.47 U.T.so that revolution 8068 crossed the central eclip‐ se line from North to South and lasted 10 min. The obscuration at this time was about 44 per cent.

Four satellite tracking centers were installed either side of the path of revolution 8068, one in Florence operated by Centro Microonde and three others in Greece located at Athens,Alexandrou‐ polis (Northern Greece) and on Crete (Southern Greece).All centers were identically equipped with two Motorola receivers working on 40 and 41 MHz with Tapetone converters,Unitron amplifiers and Sefram recorders.

For the calculation of the total electron content we used the first-order formula relating the total Faraday rotation Ω to $\int Ndh$ as indicated by BROWNE et al.(1956) :

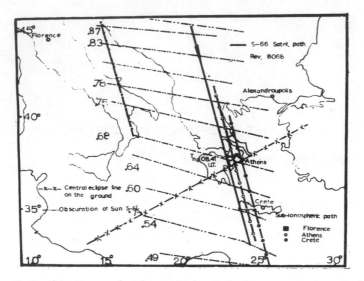

*Fig.1.Path of revolution 8068 of S-66 as was tracked from
Florence Athens and Crete,along the iso-obscuration
of Sun lines.*

$$\Omega = k/f^2 . M\int Ndh Rad$$

where : k = $2.36 \times 10^4 \times 10^3$ MKS units,
       f = satellite transmission frequency in Hz
       N = electron concentration (MKS)
       M = Bcos∂secx, B= induction of Earth's magnetic field,Wb/m^2,
       ∂ = angle between magnetic field and the ray path,
       x = zenith angle of the satellite.
Observations on 40 and 41 MHz satellite signals have enabled us to
determine Ω using the differential Faraday method.

Because all the passages used were close to the receiving
centers,it was estimated,following YEH(1956),that the correction
due to second-order effects did not exceed 5 per cent of the first
order results.

Observations during eclipse time

Table 1 gives the times of nearest approach of S-66 satellite
revolutions over Florence,Athens and Crete,during the eclipse day
(May 20) and also during the adjacent days 18th and 21st May.In the
following, we will use data only from Florence,Athens and Crete
because of the failure of the Alexandroupolis equipment,

Fig.2 shows the general variations of the total electron content vs.subionospheric latitudes along the path of revolution 8068

TABLE   1

Date	Time of near approach		
	Florence	Athens	Crete
18 May	09.29	09.31	09.32
20 May(eclipse day)	08.37	08.39	08.40
21 May	09.06	09.07	09.08

at 08.40 on the eclipse day,measured from the above three stations.
In plotting these curves,it was assumed that :

(a) differences in ionospheric area in which electron content
    was measured due to longitude variation may be neglected

(b) the percentage of obscuration of the solar disc for every
    subionospheric latitude,was very approximately the same.

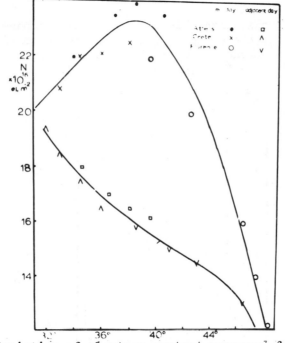

Fig. 2. Variation of electron content measured from the three
stations VS Sub-ionospheric latitude.

Fig.2 must be considered as a first approximation,giving a
general idea of how the total electron content varied with sub-ionospheric latitudes.Comparing these variations of electron content
with the variations of the mean values,during the control days
(Table I) we notice an increase of electron content and also that
the horizontal gradient changes in slope and sign at a sub-iono-

spheric latitude of 38.5°.

    For a more detailed analysis of data from the Athens station,
we used as a control curve the mean values of electron concentration computed from results obtained during the 10 days before and
after the eclipse day.For this purpose we used passages which
crossed the same sub-ionospheric latitudes at times within one hour
of the time when the satellite crossed the central line during the
eclipse day.

    The mean value of the electron concentration computed from
these recordings is shown in Fig.3. The typical spread is indicated and we see that on the eclipse the values exceeded the mean
control value: by 33 per cent.There is thus an increase of electron
content during the eclipse,instead of a decrease as might be expected. Revolution 8068 took place well after the eclipse first
contact between 08.34-08.44 U.T. and lasted 10 min.

    As a first approximation,this period of time compared with
the total duration of the eclipse of 180 min,may be considered as
instantaneous.Figure 3 then shows conditions existing in the medium between satellite and Earth,effectively at 08.40 U.T.

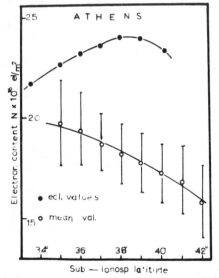

*Fig.3.Values of electron content VS sub-ionospheric
       latitude measured on eclipse and on control (0)
       days.*

    A similar maximum was observed in foF2 measured from ionograms,
obtained by Athens,during the same period.We can see (Fig.4) that
at 0840,foF2 had just passed a maximum.This maximum was reported

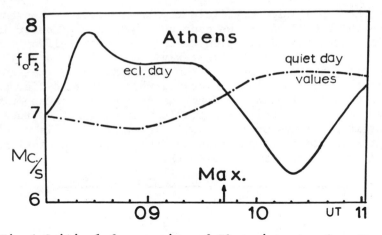

Fig. 1. *Critical frequencies of F2 region over Scaramanga station (near Athens) during eclipse and control day.*

and discussed in a previous paper by ANASTASSIADIS and MORAITIS (1968).

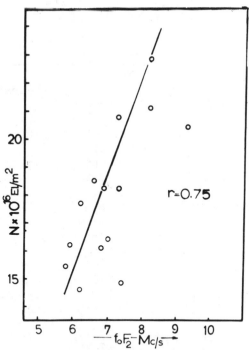

Fig. 5. *Correlation between electron content and foF2.*

Since the correlation coefficient between electron content
and foF2 is about 0.75,fig.5. we conclude that the increases in
electron content and in foF2 value during the first half of the
eclipse can be explained in the same way. An examination of the
electron content,along the path of the satellite shows that increases started to be observable at Florence,which is rather far from
the central eclipse line in the F-region.As the sub-ionospheric
latitude decreases the horizontal gradient of electron content decreases but it is always positive until latitude 38°.5 is reached
and there the gradient changes sign and the electron content continues to decrease further to the South.

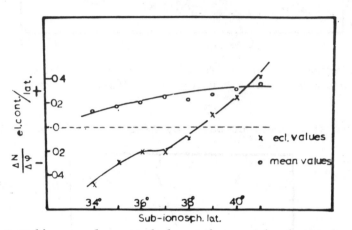

Fig.6.*Eclipse and control day values of horizontal electron*
content gradient vs.sub-ionospheric latitude.

Figure 6.shows this characteristic change in the sign of the
gradient caused by the eclipse,in comparison with the control curve.
The control curve shows the mean gradient of electron content computed from the results of the control days.

Observations]2 hours after eclipse time

As it has been shown,above the solar eclipse had a strong influence upon the electron content of the ionosphere,computed during
the eclipse.

At the same day,May 20,but twelve hours after the eclipse's
time we computed the electron content using signals from the same
satellites S-66.
The variation of electron content along the sub-ionospheric path is
compared with the corresponding variations of e.c. at the previous and next day of the eclipse and at the same hour.

At the 19th and 21st May this variation as shown in fig.7

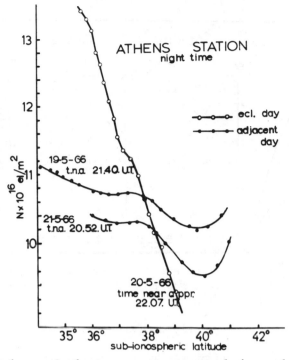

Fig.7.*Values of electron content vs.sub-ionospheric latitude
measured on eclipse day(O) and on adjacent (.) day.*

is exactly the same, but during the eclipse day this variation is
greatly different.

        This difference cannot be explained by the time or by the longi-
tudinal effect because the time of near approach and the sub-satel-
lite path are almost the same for the three passages of satellite
S-66.This difference or the horizontal gradient of electron content,
$\Delta N/\Delta \Phi$ derived from the fig.7,changes during the eclipse day.

        Consequently this is due to the effect of solar eclipse,which
still remains even 12 hours after the end of the eclipse

        The influence of the solar obscuration on the total
        electron content

        To study the influence of the percentage obscuration on the
total electron content,we computed the ratio S/S_o of the obscured
area of the solar disc to the total, as a function of latitude.For
this purpose a program of the Astronomical Institute of the National Observatory of Athens was used.For each recording center the
geometry of the eclipse and also S/S_o is different and the influence

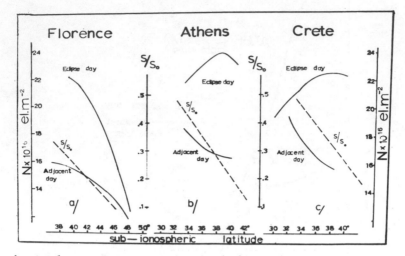

Fig.8.*Electron content N vs sub-ionospheric latitude*
eclipse and adjacent day compared with value of obs-
curation factor S/S_O

due to obscuration is consequently different.

The latitude variations of electron content, and of S/S_o during the eclipse and on adjacent control days are shown in Fig.8.

Figure 8(a) for Florence shows that for 10 per cent obscuration there was an increase of the total electron content which
continued up to 25 per cent obscuration. Thus for sub-ionospheric
latitudes 49-39° and less than 25 per cent obscuration, we observed
only an increase of electron content.

Figure 8(b) for Athens shows the same behaviour while the
obscuration was less than 25 per cent, but when the obscuration was
greater than 25 per cent there was a decrease of electron content
We can then assume that from this amount of obscuration, a different mechanism start to operate related to the decrease of the
ionizing agent. Electron content continued to decrease progressively up to the extreme southern limit for reliable measurements from
Athens.

Figure 8(c) shows the curves appropriate to Crete. This center
is also the closest to the central eclipse line in the F-region.
The electron content starts to decrease after S/S_o has reached
the value 0.25 and continues to decrease progressively as the percentage of obscuration increases.

We conclude from Fig.8 that the expected reduction of the io-

nisation occurred well after the eclipse first contact and only
when the obscuration exceeded 25 per cent.

Discussion

All values of electron content measured through the satellite
technique between 42-35°N latitude during the first half of the
annular solar eclipse of May 20,1966, were increased in comparison
with values observed from the same latitudes during control days.
Moreover the increased values reached a maximum at 38°N.

The above results indicate a competing action between the mechanism causing electron content increases and the mechanism causing reduction of electron content.

When the percentage obscuration is less than 25 per cent the
factor causing increase of electron content seems to be predominant
and when obscuration is greater than the effect of the agent redu-
cing ionization predominates.foF2,measured along the central eclipse path, was formed to behave in a similar way.

Acknowledgements -We are indebted to the Centro Microonde satellite staff for their kind collaboration in communicating eclipse data
and to the Astronomical Institute of the National Observatory of
Athens for the computation of occultation percentage of the solar
disc.

REFERENCES

KLOBUCHAR S.and WHITNEY H.   *1964 Trans Am.geophys.Un.45,350*
HAWARD H.,LUSIGNAN B.,YOH P.   *1964 J.goephys.Res.69,540*
   and ESHLEMAN V.R.
BROWNE I.,EVANS J.,HARGREAVES J.   *1956 Proc.phys.Soc.(B) 69,901.*
   and MURRAY W.
YEH K.   *1960 J.geophys.Res.65,2548*
ANASTASSIADIS M.and MORAITIS   *1968 J.Atmosph.Terr.Phys.30,1471*

IONOSPHERIC ECLIPSE EFFECTS

RECENT IONOSPHERIC E AND F REGION MEASUREMENTS DURING

SOLAR ECLIPSES, AND THEIR INTERPRETATION

J.O. Thomas[1] and M.J. Rycroft[2]

[1] Imperial College, University of London, England

[2] Physics Department, University of Southampton, England

ABSTRACT

During a solar eclipse, the electron density at a particular altitude in the ionosphere responds to the changing nature and intensity of radiation incident on the upper atmosphere. Theoretical expressions are presented which relate the changes of electron density which occur with time during an eclipse to the production of ionization (by solar photospheric / coronal radiations and other mechanisms), and to the rates of electron loss by the relevant chemical processes of recombination.

It is suggested that, instead of estimating electron loss coefficients from eclipse measurements as in the past, it is more useful to assume these are known from laboratory and other measurements and to deduce estimates of the magnitudes of the contributions of various sources of ionizing radiation responsible for the production of the plasma.

Measurements made during the eclipse on the morning of 15th February, 1961, at several observatories in Europe and the U.S.S.R. are compared. This eclipse is important because it appears to be the first in which synoptic ionospheric measurements were made at a number of observatories so that the method of superposed epochs can be used. It is interesting also because the eclipse occurred near sunrise and, at some observatories, rose eclipsed. A detailed study of this eclipse is in progress. The full results will be reported elsewhere.

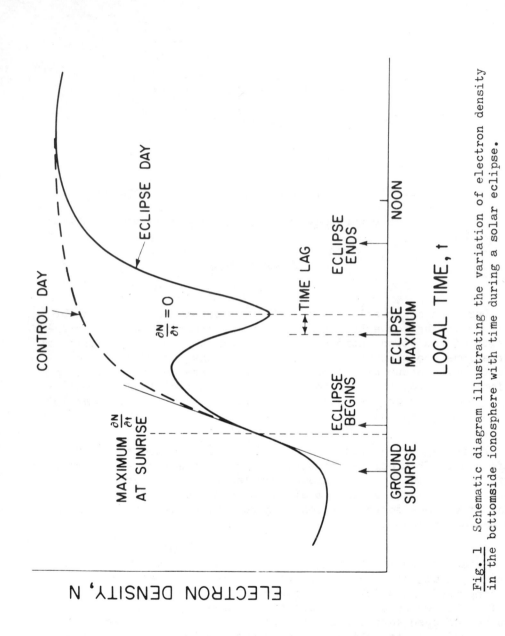

<u>Fig. 1</u> Schematic diagram illustrating the variation of electron density in the bottomside ionosphere with time during a solar eclipse.

INTRODUCTION

It has long been recognized that observations of the changing
ionospheric electron density, N, during a solar eclipse, can aid
the study of the physical processes operating in the ionosphere.
This is because the fraction of the solar disc illuminating the
atmosphere at any height, h, above any point on the earth's surface
is, at these times, known with the great precision and accuracy
that are characteristic of astronomical computations. Hence, assuming the extreme ultra-violet radiation to be emitted uniformly
from the solar disc, the rate of production of electrons, q, by
photoionization of a model atmosphere can be calculated according
to the theory of Chapman (1931). These electrons may then be lost
by a recombination-type process, by an attachment-type process, or
by motion out of the region where they were produced.

Recently, the rates of chemical reactions of aeronomic interest,
for example the dissociative recombination rate of nitric oxide and
of molecular oxygen ions, have been measured accurately in the
laboratory, and have been collected and published in the DASA Reaction Rate Handbook (1967). Application of these values to ionospheric electron densities can, in principle, lead to information
concerning radiation which is emitted non-uniformly over the solar
disc and from beyond the solar disc. These radiations come from
active areas on the sun itself, for example, from plages and sunspots. In addition, the hot solar corona radiates X-rays, the intensities of which are particularly large from active coronal condensations.

Complete bottomside electron density profiles above ionosondes
can be obtained at various times, t, by appropriate reduction of
the ionograms, enabling N(h,t) curves to be drawn. A schematic
diagram illustrating a typical temporal variation of electron
density at a particular altitude is shown for a morning eclipse in
Figure 1. Examples of the actual variations of electron density
with height and time during eclipses have been published in the
volume edited by Beynon and Brown (1956). In a number of papers,
experimental and theoretical eclipse results are compared and electron production and loss rates are deduced. When following this
procedure for a single station, difficulties are often encountered,
since the number of unknown variables is greater than the number of
known parameters. Thus, in order to remove ambiguities and to make
the problem tractable, simplifying assumptions have to be introduced.
The conclusions reached after this procedure are often quite varied:
for instance, Rawer (1956) finds that the recombination rate may
differ on the eclipse and control days; Minnis (1956) finds that
radiation from the uneclipsed solar corona causes appreciable electron production; Ratcliffe (1956), Nesterov and Taubenheim (1963)
and others deduce that the Western limb of the sun is brighter than
the remainder. Alternatively, Burkard (1956) reports that the

atmospheric temperature increases rapidly after the eclipse; Evans
(1965) also observes large changes in the plasma temperature.
Thomas and Robbins (1956) and Danilkin (1963) have investigated
possible changes in the F region due to vertical movements.

It is evident that there are many effects to be investigated
in any attempt to fit theoretical N(h,t) curves to those observed
in an eclipse. Not only does the magnitude of the calculated electron density perturbation have to follow that observed at each and
every height, but also the calculated time lag (see Figure 1)
between optical totality and minimum electron density has to agree
with that observed at all heights. Because of the decrease of
electron loss coefficient with increasing height in the ionosphere,
this time lag, sometimes known as the "sluggishness" of the ionosphere, should increase with increasing height. However, this time
lag will be reduced if there is an appreciable flux of ionizing
radiation from the corona above the West limb of the sun which becomes uneclipsed at the time of maximum optical obscuration.

It is also necessary that good information is available on the
control days, that is to say, it is important to know how the ionosphere would have behaved had there not been an eclipse. Control
day measurements can yield information on the net electron production rates at different heights due to photospheric, chromospheric
and coronal radiations, as well as the rate of production of ionization due to other mechanisms.

In this paper, examples of N(h,t) curves deduced from observations made at four stations in Europe on the occasion of the solar
eclipse of 15th February, 1961, are presented together with the
basic theoretical relationships that are necessary for their interpretation.

THEORY

Theoretical considerations of phenomena giving rise to ionospheric electron density variations are outlined in this section.
The continuity equation for electrons can be written (Ratcliffe,
1960)

$$\dot{N}(h,t) = Q - L - M . \tag{1}$$

Here, Q and L represent production and loss terms respectively, and
the last term allows for the effect of movements of ionization.
The quantity $\dot{N}$ represents the partial derivative of N with respect
to time. Simple expressions for each term will be stated in turn.

Chapman (1931) considered the flux of monochromatic ionizing
radiation incident at an angle χ on an isothermal single-constituent

atmosphere of scale height H. The rate at which electrons are produced by photoionization at height h is given by

$$q(h,\chi) = q(y) = q_o \exp\angle 1 - y - Ch(\chi,H)\exp(-y)\angle 7,\qquad (2)$$

with $y = \angle h - h_m(\chi = 0)\angle 7 / H$, and with $h_m(\chi = 0)$ the height at which
the production rate is maximum for an overhead sun, the latter being
q_o. It should be remembered that the actual height of maximum production rate decreases during the morning hours, as the solar zenith
angle χ decreases appreciably below 90°. For values of $\chi < 75°$, the
Chapman function $Ch(\chi,H)$ is approximately equal to sec χ; for $\chi > 75°$,
near sunrise or sunset, values have been tabulated by Wilkes (1954).
The solar zenith angle may be expressed as a function of time, in
terms of the local hour angle h' of the sun measured westwards from
apparent noon (Davies, 1965),

$$\cos\chi = \sin\emptyset \sin\delta + \cos\emptyset \cos\delta \cos h',\qquad (3)$$

with $\emptyset$ being the geographic latitude and δ the solar declination.

During an eclipse, the production function q is modulated by a
function u(t) which accounts for the obscuration of EUV in the wavelength range between 170 Å and 1216 Å. This radiation is emitted by
the photosphere and chromosphere, uniformly over the solar disc.
Functions representing the effects of irregular sources on the solar
disc, for example plages, and X-radiation (< 170 Å) from the corona
are also convolved with the obscuration function. It is X-radiation
having wavelengths between 10 Å and 170 Å, and EUV radiation between
the ionization limits of atomic oxygen (911 Å) and molecular oxygen
(1027 Å) that is responsible for E region ionization. The F layers
are produced by EUV radiation between 170 Å and 911 Å, including
the important helium lines at 304 Å and 584 Å (see for example,
Risbeth, 1968).

Apart from ionization by direct solar radiation, there are
other possible sources of E and F region ionization which may have
to be taken into consideration. For example,

(a) Scattered radiation, i.e. solar radiation scattered by
the earth's atmosphere into the eclipsed region. Fortunately, since
the moon has no atmosphere, radiation scattered at large distances
(~ 60 earth radii) need not be considered.

(b) Geocoronal radiation: Lyman α and β radiation from the
earth's hydrogen envelope.

(c) Interplanetary medium: Lyman α and helium (584 Å and
304 Å) radiations from the interplanetary gas (Tohmatsu, 1969).

(d) Interstellar medium: fast hydrogen atoms, formed by the charge transfer reaction between solar wind protons and interstellar hydrogen atoms, could produce appreciable ionization on collision with atmospheric matter (Blum, 1969).

(e) Meteors: excess E region ionization is produced during periods of enhanced meteor activity.

(f) Stellar radiation: the bright X-ray star, Scorpius XR1, appreciably affects D region ionization when making a transit (Mitra and Rao, private communication, 1969). The effects of other bright X-ray and EUV stars should be considered. The integrated effect of radiation from the other 10^{11} stars in the Milky Way could be important.

(g) Magnetospheric charged particles: excess ionospheric ionization is associated with the greater precipitation of particles during periods of enhanced magnetic activity. Photoelectrons from the magnetic conjugate area could produce some ionization, but their main effect seems to be to heat the neutral atmosphere.

It is expected that the sum of these effects is small compared with the effect of direct radiations from the sun, and can be represented - to first order - by a constant rate of production of ionization at each altitude $q_1(h)$.

The dominant processes by which electrons are removed from the lower ionosphere are three-body recombination, or ion-atom charge exchange followed by dissociative recombination, the rate coefficient of the latter reaction being α. In each case, an electron effectively recombines with a positive ion of, for example, nitric oxide. The loss rate is given by

$$L_R = \alpha N^2, \tag{4}$$

α being the recombination coefficient, $\sim 2 \times 10^{-7}$ cm^3 sec^{-1}. Laboratory measurements indicate that α might increase somewhat, due to temperature reductions during the eclipse; such changes will not be considered here.

However, in the upper F region, where the neutral gas density is much smaller than in the D region, the electron loss rate is determined by the ion-atom charge transfer rate, and is given by

$$L_A = \beta N. \tag{5}$$

The attachment coefficient, β, decreases with increasing height, with a scale height equal to that of the density of the pertinent neutral gas; at 180 km altitude its value is $\sim 10^{-3}$ sec^{-1}. In the upper E and Fl regions of the ionosphere, L_A and L_R are approximately equal; with both processes operating, the net electron loss rate is given by

$$L_N = \alpha\beta N^2 / (\alpha N + \beta).$$ (6)

In the F2 region, motions of the plasma brought about by diffusion in the gravitational, electric and geomagnetic fields are important. When variations of ionospheric parameters in the vertical direction are much larger than those in the horizontal plane,

$$M = \text{div}(N\underline{v}) \simeq \frac{\partial}{\partial h} (Nv_h),$$ (7)

where v_h is the vertical velocity. Although this approximation is valid at times, it is possible that it is not so during a solar eclipse. For example, the diffusion coefficient, which is proportional to the ratio of the sum of the electron and ion temperatures to the ion-neutral collision frequency, could change during the eclipse. Besides diffusion, there are other forces acting on the plasma which tend to move the plasma. The gravitational attractions of the sun and the moon, the origin of tidal effects, reinforce each other during an eclipse. An electrodynamic drift, having an upwards component, is produced in the Northern hemisphere by an Eastward directed electric field. Neutral particle winds, which blow from hot regions to cold at almost constant altitudes, tend to drag the ions with them. In the morning hours, this mechanism would push plasma down lines of force of the geomagnetic field (Kohl and King, 1967). This wind system might be perturbed as temperatures might change somewhat during the eclipse. It is possible that atmospheric gravity waves could be excited during the eclipse; these might give rise to detectable electron density perturbations associated with the eclipse but beyond the eclipse path.

For the situation during a solar eclipse, equation (1) may thus be written generally as

$$\dot{N}(h,t) = \underline{/}u(h,t)\ f(t) + g(t)\underline{7}\ q(h,\chi) + q_1(h)$$

$$- \alpha(h)\beta(h)N^2(h,t) / \underline{/}\alpha(h)N(h,t) + \beta(h)\underline{7}$$

$$- \text{div}\ \underline{/}N(h,\ t)\ \underline{v}(h,t)\underline{7}.$$ (8)

In this equation $u(h,t)$ is the fraction of uneclipsed solar disc, $f(t)$ is a function accounting for the effects of radiation from localized sources on the sun, including limb brightening, and $g(t)$ allows for the fluctuating coronal source. Essentially steady, non-solar, sources of ionization of the atmosphere are represented by $q_1(h)$. For simplicity, the loss coefficients α and β are assumed

to be independent of time; that is, temperature changes during the
eclipse do not appreciably change the rates of electron recombination. Also for simplicity, the effect of movements will be neglected; such an approximation is reasonable at E and Fl region altitudes. Considering one altitude and remembering that χ is a function of time t, equation (8) becomes

$$\dot{N}(t) = \underline{/}u(t)f(t) + g(t)\underline{/}q(t) + q_1 - \alpha\beta N^2(t) / \underline{/}\alpha N(t) + \beta\underline{/}.$$

$$(9)$$

For steady localized photospheric sources and for steady coronal
radiation, then, on the control days for which u = 1, or on the
eclipse day before the eclipse commences,

$$\underline{/}u(t)f(t) + g(t)\underline{/} = k(say) = 1 \times f + g \gtrsim 1. \qquad (10)$$

The multiplying factor, k, whose value typically lies between 1.1
and 2, accounts for the extra ionization production rate due to
active solar areas and to the corona, above and beyond that due to
the uniformly radiating disc, namely q_o in equation (2).

   Equations (9) and (10) can be used to analyse observations at
each particular altitude which have the same form of N(t) curve as
that presented in Figure 1. It is of particular interest to note
that after sunrise $\dot{N}$ reaches a maximum value. After the eclipse
starts, $\dot{N}$ passes through zero and becomes negative, passes through
zero again some time after the maximum obscuration effect of the
eclipse, and then increases rapidly until the electron density
reaches its normal value. At the first of the three points of in-
flexion on the N(t) curve shown in Figure 1, where $\dot{N}$ is an extremum,
a maximum, we can equate the partial second derivative with respect
to time, $\ddot{N}$, to zero:

$$\ddot{N} = \underline{/}q\dot{k} + k\dot{q} - r\dot{N}\underline{/}_{ext\ 1} = 0 \qquad , \qquad (11)$$

where $r = \dfrac{\alpha\beta N(\alpha N + 2\beta)}{(\alpha N + \beta)^2}$,

thus $r = \underline{/}(q\dot{k} + k\dot{q}) / \dot{N}\underline{/}_{ext\ 1} .$ $\qquad (12)$

Together with equation (9), expressed in the form

$$\dot{N} = \underline{/}kq + q_1 - \dfrac{\alpha\beta N^2}{(\alpha N + \beta)}\underline{/}_{ext\ 1} \qquad , \qquad (13)$$

we thus have two equations, (12) and (13), for the three unknowns

$\overset{\bullet}{k}$, k, and q_1, the quantities q(t), N and $\overset{\bullet}{N}$ being evaluated at the
time when extremum 1 is reached. Similar equations can be obtained
at the other two extremum values of $\overset{\bullet}{N}$, giving six equations for the
unknowns f, g, $\overset{\bullet}{f}$, $\overset{\bullet}{g}$ and q_1. Thus assuming that each of the functions
f and g may be approximated by a simple two parameter function during
the interval (~ 2 hours) between extrema 1 and 3, and assuming a
Chapman layer in the ionosphere, observations of the electron density
and its time rate of change can lead to information on production
rates at heights where both recombination-like and attachment-like
mechanisms of electron loss operate, that is, over an appreciable
height interval from ~ 130 km to ~ 230 km. If these three independent measurements made during the eclipse lead to consistent
results at each height, the approximations made can be considered
to be valid.

Additional complications are introduced by possible temperature
changes with time during the eclipse, which alter both the neutral
gas scale heights and the electron loss coefficients.

Other special cases of equation (9) can be investigated simply.
For example $\overset{\bullet}{N} = 0$, at some time τ_1 during the waxing part of the
eclipse effect and at another time τ_2 after the maximum eclipse of
the sun,

$$\alpha \beta N^2(\tau) \, / \, \underline{/\alpha N(\tau) + \beta \underline{/}} = \underline{/u(\tau) \, f(\tau) + g(\tau) \underline{/} q(\tau)} + q_1, \qquad (14)$$

again assuming movements to be unimportant. Knowing α and β and
with an estimate of the value of q_1, two values of f and g and of
$\overset{\bullet}{f}$ and $\overset{\bullet}{g}$ could be obtained. As a check, the values of $\overset{\bullet}{N}$ must be
negative and positive respectively.

Another special case of equation (9) occurs for u(t) equal to,
or close to, zero. Then the ionospheric effects of bright areas on
the sun are minimized, so g(t) may be assessed independently of
f(t). If these values are mutually inconsistent, significant movements of ionization must be taking place; these are expected to be
more noticeable higher in the ionosphere.

ECLIPSE OF 15th FEBRUARY 1961

There are two features of the solar eclipse of 15th February,
1961, which make its ionospheric effects especially worthy of study.
Firstly, it took place near sunrise, on a day of moderate solar
activity. Secondly, its path fell across the continent of Europe
and the U.S.S.R. where there are many stations recording ionograms
typical of mid-latitude conditions on a routine basis. Wherever
possible, ionograms were recorded at approximately five minute
intervals within the one or two hour period preceding and following

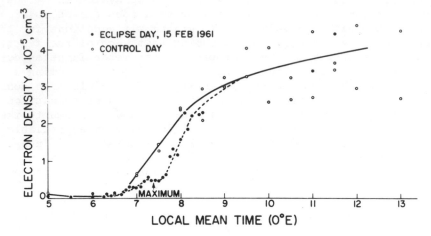

Fig. 2 Temporal variation of electron density at 180 km altitude, above Ebro. It should be noted that on the eclipse day (dark circles) there are only two points after 08.30 U.T. Nearly half the points are not as reliable as the others because of absorption, sporadic E, spread F, etc.

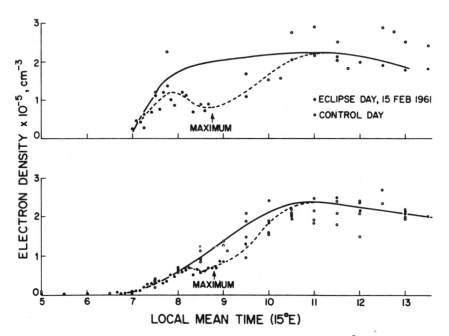

Fig. 3 Temporal variation of electron density at 180 km altitude, above Graz (upper part) and Lindau (lower part).

the eclipse maximum. On the control days, ionograms were, in general, taken every half-hour or hour throughout the day. This observational schedule was not strictly adhered to at all the stations. Also, strong absorption of the ionosonde radio waves or marked reflection from a sporadic E layer sometimes took place, in which case N(h) profiles were less reliable.

Ionograms recorded at the stations of Athens, Ebro, Graz and Lindau were communicated to one of the authors (J.O.T.) at the Cavendish Laboratory, University of Cambridge, where they were reduced to N(h) profiles by the technique described by Thomas, Haselgrove and Robbins (1958), and by Thomas and Vickers (1959).

The temporal variations of F1 region electron density (between $< 10^4$ cm^{-3} and $\sim 5 \times 10^5$ cm^{-3}) are plotted for h = 180 km in Figures 2, 3 and 4. Individual data points on the eclipse day, computed at five minute intervals, are shown as dark circles. Some scatter between neighbouring values is apparent. This may be due partly to ionospheric irregularities and partly to errors associated with converting the ionograms to electron density profiles, for example in the determination of the altitude of the minimum detectable electron density. The broken curve is drawn through the points observed on the eclipse day. It is clear that it is difficult to make an accurate estimate of the time lag between the eclipse maximum and the minimum electron density.

There is much scatter in the control day data recorded on 13th and 14th February (Figure 2, open circles), through which the continuous curve is drawn. Electron density values at Ebro on 14th February are slightly lower than those on 13th between 07.00 and 09.00 U.T., but rather higher, by about 30%, between 09.00 and 13.00. It is impossible to know whether 15th February, the eclipse day, would have more resembled 13th or 14th February had there not been an eclipse.

Referring to Figure 3, there is a gap in the Graz data points between 07.40 and 08.30 U.T. (08.40 and 09.30 Local Mean Time) at a crucial stage in the development of the eclipse effect. Nothing can be said about the time of minimum electron density in relation to the time of maximum eclipse obscuration (90%). Again on the control days (13th and 14th February), the electron densities are $\sim 50\%$ higher on the 13th between 07.30 and 13.00 L.M.T.

At Lindau, eclipse-day data are sparse after 09.00 L.M.T. However the minimum electron density is reached about 15 minutes before the maximum obscuration (only 84%) of the solar disc - an unexpected effect. Regarding control day data at Lindau, electron density values are $\sim 20\%$ higher between 08.00 and 11.00 L.M.T. on the 13th than on the 14th. Other control day data are contributed by measurements made on 12th and 20th February, 1961.

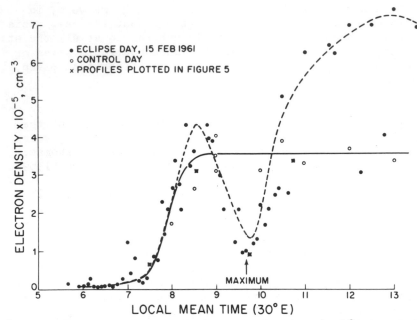

Fig. 4   Temporal variation of electron density at 180 km altitude
above Athens.   There is a marked difference between the behaviours
on the control days and after 10.00 L.M.T. on the eclipse day.

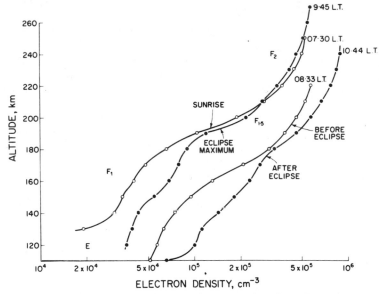

Fig. 5   Altitudinal distribution of ionospheric electron density,
on a logarithmic scale, above Athens at four times during the solar
eclipse of 15th February, 1961.   Values at 180 km altitude are
shown as crosses in Figure 4.

Figure 4 shows an excellent example of F1 region ionospheric response during a solar eclipse. At 180 km altitude above Athens there is a time lag of ∿ 5 minutes between maximum obscuration (98%) and minimum electron density. It is noted that the observed values are rather small about half an hour before the minimum electron density is reached. Ilias and Anastassiadis (1964) have shown that this effect is explained by enhanced radiation from an active region near the West limb of the sun contributing to f(t) in equation (9). This active region is covered during the initial phase of the eclipse. An unusually large electron density value is observed at 08.47 L.M.T. (06.47 U.T.). Slightly larger values are noted at the three other stations at either 06.45 or 06.50 U.T. These could be due to an enhanced flux of ionizing radiation from a solar flare reported, by optical methods, between 06.40 and 07.00 U.T. Radio emission at 108 MHz from the East limb of the sun between 06.55 and 07.18 U.T. is also discussed by Ilias and Anastassiadis (1964). These phenomena suggest that f in equation (9) can be a complicated function of time t.

Danilkin (1963) has presented useful N(h,t) curves deduced from ionograms recorded at Simferopol', Ukraine. Similar curves for other stations have been reported in the scientific literature.

Changes in the detailed structure of the ionosphere as the eclipse progresses are now discussed briefly. Four electron densit profiles above Athens, at approximately hourly intervals, are shown in Figure 5. The lefthand curve, shown by open circles, is taken a 07.30 local time, only 15 minutes after ground sunrise. The E region is not yet formed, in contrast with the F1 and F2 regions. At 08.33 L.T., the E region is well developed and the electron density at 180 km is nearly five times greater than an hour earlier: the eclipse is just starting – the obscuration at 180 km is only 2%. One hour and twelve minutes later, the electron density at all altitudes at eclipse maximum (98%) is reduced by a factor $\gtrsim$ 2. The F1 layer is pronounced, and an F1.5 layer has developed at ∿ 200 km altitude. Near the end of the eclipse (11% obscuration) electron densities, shown by dark circles, have recovered to a value greater than their values before the eclipse commenced, by an altitude dependent factor which is smallest at 180 km altitude.

A detailed analysis of the observations, described in a preliminary way here, is in progress using a theoretical approach along the lines suggested in the first part of this paper. The results will be published elsewhere.

References

Beynon, W.J.G., and G.M. Brown, ed., Solar eclipses and the iono-
    sphere (abbreviated S.E.I.), Pergamon Press, London, 1956.

Blum, W.P., Paper A.3.15, COSPAR, Prague, 1969.

Burkhard, O., S.E.I., Pergamon Press, London, 69, 1956.

Chapman, S., Proc. Phys. Soc., 43, 26 and 484, 1931.

Danilkin, N.P., Geomagnetism and Aeronomy (English Translation),
    3, 388, 1963.

DASA Reaction Rate Handbook, 1967.

Davies, K., Ionospheric radio propagation, National Bureau of
    Standards Monograph, 80, 1965.

Evans, J., J. Geophys. Res., 70, 131, 1965.

Ilias, D., and M. Anastassiadis, Electron density distribution in
    ionosphere and exosphere, ed. E. Thrane, North-Holland,
    Amsterdam, 170, 1964.

Kohl, H., and J.W. King, J. Atmos. Terr. Phys., 29, 1045, 1967.

Minnis, C.M., S.E.I., Pergamon Press, London, 158, 1956.

Nestorov, G., and J. Taubenheim, Geomagnetism and Aeronomy (English
    Translation), 3, 224, 1963.

Ratcliffe, J.A., S.E.I., Pergamon Press, London, 1, 1956.

Ratcliffe, J.A., ed., Physics of the upper atmosphere, Academic
    Press, New York and London, 1960.

Rawer, K., S.E.I., Pergamon Press, London, 198, 1956.

Rishbeth, H., Rev. Geophys., 6, 33, 1968.

Thomas, J.O., and A.R. Robbins, S.E.I., Pergamon Press, London, 94,
    1956.

Thomas, J.O., J. Haselgrove and A.R. Robbins, J. Atmos. Terr. Phys.,
    12, 46, 1958.

Thomas, J.O., and M.D. Vickers, D.S.I.R. Radio Research, Special
    Report No. 28, H.M.S.O., London, 1959.

Tohmatsu, T., Paper A.3.7., COSPAR, Prague, 1969.

Wilkes, M.V., Proc. Phys. Soc., $\underline{B67}$, 314, 1954.

THE ANNULAR SOLAR ECLIPSE ON MAY 20 1966 AND THE IONOSPHERE

Michael Anastassiades

University of Athens

The annular solar eclipse over the North part of Africa and the East part of Europe on May 20, 1966, was observed by a number of ionosonds, located along or near the central eclipse path.

The Ionospheric Institute, of the National Observatory of Athens operated two vertical ionosonds located in the central eclipse line of the E layer, or in its vicinity. Scaramanga sounder with 30 km NW of the above mentioned eclipse path and Gytheion Cossor type sounder some 200 km SW of Scaramanga was located in the central eclipse path.

Both locations offered the possibility of observations of the near totality eclipse at heights of the E and F layers. Because the eclipse occurred around noon local time , the Sun was at an approximate elevation angle of 21°. Following the eclipse geometry the ionospheric layers experienced, at a given time, different degrees of obscuration of the Sun. Table I lists the time of events at different heights over the Scaramanga ionosond

T A B L E   I
Eclipse Schedule

Height (km)	1st Contact UT			Max. Eclipse			Totality	4th Contact		
	h	min	sec	h	min	sec		h	min	sec
0	08	03	17,3	09	31	47,4	0,988	11	05	00,2
100	08	04	51,8	09	31	50,1	0,978	11	05	45
200	08	04	27,3	09	31	53,4	0,968	11	06	30
300	08	04	03,8	09	31	57,4	0,958	11	07	16

Fig.1 shows the eclipse path computed at ground,100,200 and
300 km levels over Greece. Differences in time of maximum between
ground and 200 km levels, are of the order of 1 min.

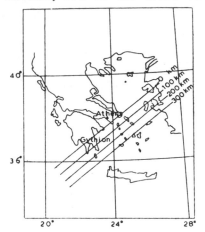

Fig.1.Eclipse path computed at ground,100,200 and 300 km levels.

Along the eclipse path several ionosonds were in operation.
Our discussion on the eclipse phenomena of May 20,1966 is based
on results from observations made by the CRI French group, on
results from the Greek network of ionosonds of the Ionospheric
Institute of the National Observatory of Athens, and on observations made by the Istanbul ionosond. These stations are listed
in Table II in the order in which maximum obscuration occurred.

T A B L E   II

Location	Longitude	Latitude	Magni.dip	Time max. obscurat. (ground level)	Obscur. % at 20
Dakar	17.26 W	14.40 N	22.47 N	08.08 UT	75%
Ouagadougou	1.40 W	12.18 N	9.49 N	08.09 UT	89%
Tamanrasset	5.31 E	22.48 N	30.26 N	08.34 UT	87%
Gytheion	22.34 E	36.45 N	51.50 N	09.28 UT	100%
Scaramanga	23.36 E	38.01 N	53.30 N	09.31 UT	97%
Istanbul	29.02 E	41.08 N	57.40 N	09.43 UT	95%

We can see from Fig.2 that Tamanrasset, Gytheion and Athens
ionosonds were in the central or almost in the central ground eclipse path, while Dakar lay to the North. The central line of the
annular solar eclipse of 20 May 1966 traversed a region of W.Africa,
Algeria,Central Mediterranean sea, Greece South to North and the
NW part of Turkey, Greece and Turkey were located in the semidura-
tion area of 90 min, Algeria in the 80 min, High Volta in the 70

min and Senegal in the 60 min.area.

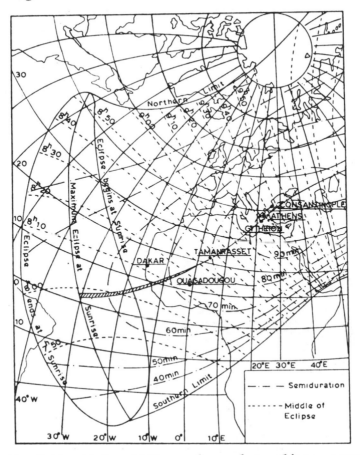

Fig.2. Path of the 20 May 1966 annular solar eclipse, over Northern Africa and Southern Europe at ground level.

In order to form a basis for comparison on the data recorded on the eclipse day it was necessary to determine the average values for the parameters of the ionosphere. Data for 10 days were therefore reduced (19-30 May) omitting 20th May, and an average of the values is calculated and plotted on Fig.3. Scaling of the values of the virtual heights and critical frequencies from the ray ionograms are accurate in 10 km and 0.1 Mc/s. Values recorded from the ionograms sometimes change in step but they are smoothed in the judgement of the reader of the ionograms in plotting curves.

No magnetic disturbances occurred during the control and the eclipse days. Only five days at the beginning and at the end of May 1966 were magnetically disturbed. Vertical sounding values were obtained every minute. Scaramanga ionosounder is equipped with a special device which makes possible the continuous recordings of several ionospheric parameters like foE, foF1, foF2, hmaxF etc.

Three active centers were present on the day of the eclipse
in the solar disc. Their distribution over its surface was such
that they were covered and uncovered in succession by the Moon as
determined by optical observations, at ground level over Athens.

Active center :	A	B	C
Beginning and end of occultation :	08.22 - 09 39	08 44 - 10 19	0928 - 11 04 UT

Fig. 3 shows Ha photos of covering and uncovering at typical
moments. Fig. 4 shows the Fraunhofer map of the Sun on the day of
the eclipse.

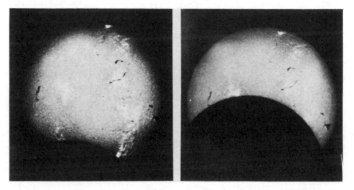

08.22 U.T. (A)   08.51 U.T. (C)

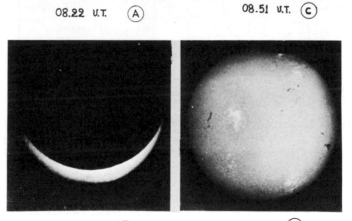

09.41 U.T. (A')   11.02 U.T. (D')

*Fig.3 Ha photos of covering and uncovering of the solar
disc during the May 20,1966 at typical moments,taken
by the Astronomical Institute of the National Observatory of Athens.*

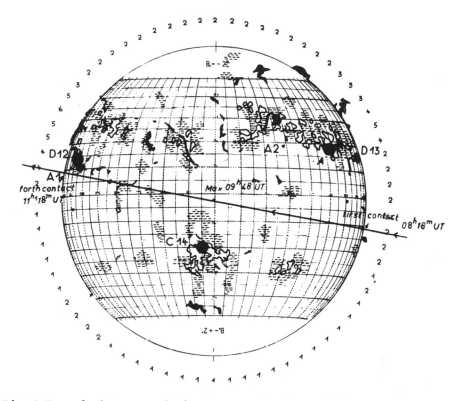

Fig.4.Fraunhofer map of the Sun on the day of the eclipse

ECLIPSE CURVES FOR DIFFERENT IONOSPHERIC LAYERS

E - region

E-region virtual height remained at 110 to 120 km throughout
the 10 control days and the day of the eclipse. At 08 03 UT (time
of first contact) a drop of foE,critical frequency of the E layer
was noted, indicating a prompt response to the eclipse effect.
Scaramanga ionosonde near Athens measured values of foE until
08 30 UT. After this moment a sporadic E layer renders impossible
any foE measurement. Occurrance of Sporadic E during May in our
latitude is rather high. In May 1966 its appearance lasted all
10 control days. No measurements were obtained from the E layer
in Gythion, because of the permanent appearance of Sporadic E.

Constantinople ionosonde measured foE throughout the eclipse
period.Fig.5 shows the part of the foE eclipse curve from Athens
and the complete foE eclipse curve from Constantinople.Athens and
Constantinople ionosonds are of the same type. It is important to
note that in both locations, foE decrease due to the eclipse star-

ted earlier than the first optical contact.

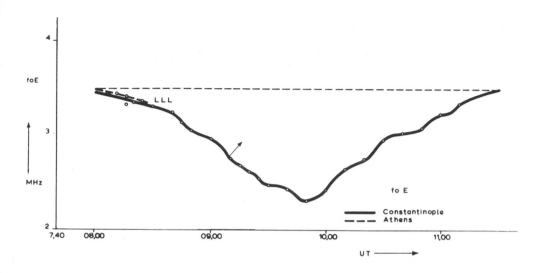

Fig.5.*Critical frequency variations in the E-layer measured by Constantinople ionosonde.Dotted line indicate some foE values measured by Athens ionosonde,before occultation by Es.*

F1 - region

F1-region measurements of the critical frequencies were made by all ionosonds across the eclipse path. The abnormal formation of an F1 layer during an eclipse is considered as a typical phenomenon. Such a formation was observed over Athens during the solar eclipse on February 15,1961. In the eclipse of May 20,1966 however and in the course of the 10-day period preceding the eclipse and the 10-day period after the eclipse an F1 layer was observed between 08 00 and 10 00 UT.

Fig. 6 shows the F1 layer eclipse curves measured in Gytheion and Athens. Both are far from being smooth. A 40% decrease is observed between control days foF1 and at eclipse maximum. The symmetry of foF1 values around each local eclipse maximum is of particular importance. The similarity also of foF1 deviations from the optical eclipse variations in both curves, cannot be due to measuring errors only.In any way they cannot be produced by the occultation of a uniform solar disc.

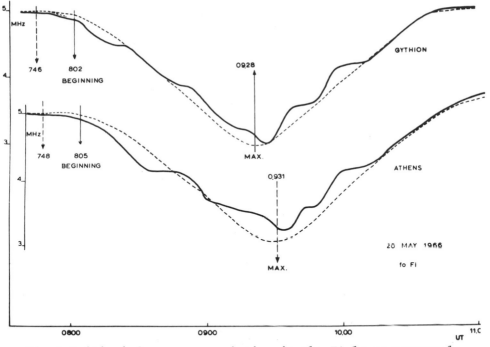

Fig.6.*Critical frequency variation in the F1 layer measured
in Scaramanga (Athens) and Gytheion.*

Three active centers were present in the solar disc during the
solar eclipse of May 20,1966. Boviatsos and myself investigated the
effect produced in different ionospheric layers by their occultation
It was shown that a hard x-ray emission associated with the above
centers was responsible for the variations in absorption observed
in the D-layer, and that absorption variations of much less importance observed in the E-layer, may be associated with the emission
of soft x-rays from part of the solar disc in the vicinity of the
above centers.

foF1 deviations from the optical eclipse variations must be
produced by the occultation of the parts of the Sun emitting soft
x-rays radiation. But the solar disc is not the exclusive source
of such a Lyman-a radiation. Also the corona is contributing,and
in examining foF1 deviation we must consider in some more detail
the occultation of the corona parts.

Fig.7 shows the distribution and the flux importance of active
centers and corona on the Sun on May 20,1966, measured by Castelli
of the Air Force Cambridge Research Lab. and Caroubalos and myself
by using radiastronomical devices working on 10 cm. in the central
eclipse path in Athens. We can see that on the West and East limb

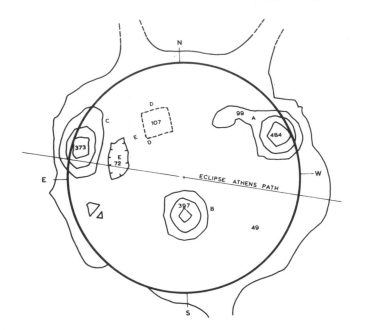

*Fig.7.Distribution and flux importance of corona brightness
and of active centers following radiastronomical
measurements on 2.980 MHz.*

of the solar disc, active centers located there, are extended to
the corona and that there are three corona prominances, one at the
West limb, the second of higher importance in the NW limb and a
third of less importance in the NE limb.

If we accept that corona occultation may produce eclipse effects in the F1 layer then we can justify the fact that foF1 values
started to decrease at 07 48 UT, which following the Athens eclipse
geometry is the time of the first contact of the Moon with the West
prominance of the corona. First contact with the solar disc is
08 05. We can then consider that the foF1 decrease between 07 48 UT,
and 08 05 is due to the West limb corona occultation, in Athens. A
similar effect was also observed at Gytheion (Fig.6).

It is interesting to follow the deviations of foF1 from the
optical variations curve,in the progress of the eclipse. Following
the Athens eclipse geometry at 08 34 the West limb corona and the
active center denoted by A is covered and we observe the maximum of
foF1 decrease. The deviation is high and we may assume that this
West part has an important contribution to the total solar radiation
Between 08 40 and 09 15 UT no deviation was observed. But at 09 20
UT foF1 values are increased and an upward deviation is observed.

This may be explained by the antagonistic action between the uncovering of the West part of the corona and the covering of parts
of the solar disc in the vicinity of active center C.

The maximum occulation of the solar disc in Athens was at
09 31. The foF1 maximum decrease occurred 4 minutes later, and foF1
was higher than the expected values. This is obviously due to the
corona effect which is acting in its totality at this moment.

foF1 variations after the maximum of occultation support the
assumption made on the corona contribution. For a large part of
the uncovering eclipse period of the Sun foF1 values are higher
then those expected from the variations of the optical curve.The
uncovering of the very active West part during the covering of the
less active East part, the quick occultation and just after uncovering of part of the NE corona prominance, may be followed in the
deviations of foF1 values.

It may then be concluded that in spite of the general idea that
the ionosphere is not an adequate means to investigate variations of
the solar flux the above behavior of the F1 layer observed proves
the contrary.

In order to estimate the part of contribution of the corona
to the total radiation, we may compare the total air of the surface determined by the contour of the corona indicated in Fig.7 to
the air of the surface of the solar disc, considered as uniform.
We found that the corona contribution is 30% of the total radiation.
The remaining 70% is due to the uniform solar disc. This amount of
70% must be reduced even more in order to take in account the contribution from the three active centers. If we consider the rather
minor absorption effects due to these centers in the E layer, their
contribution to the ionizing radiation must not exceed 5%.

They are two alternatives for the aeronomical estimation of
the ionizing radiation of the eclipsed Sun. If we consider the recombination coefficient as constant we have to increase the effective surface of the solar disc.Otherwise if we accept that the
ionizing radiation is proportional to the uncovered part of the
solar disc we have to accept that the recombination coefficient
changes simultaneously with the change of the ionizing radiation.

From these two alternatives we prefer the first: for a constant
recombination coefficient and an increased solar surface. From the
time delay between the optical and the foF1 minimum curves we may
estimate that the recombination coefficient is high. By using the
Dominici method, we found a of the order of 10^{-7} wich is a pret-
ty high value. We keep this value constant and we increase the solar surface. Then we assume in a first approximation that the
electron production rate q is proportional to the electron density

$$q \simeq aN^2$$

The variations of the ionizing radiation throughout the eclipse may be expressed by q/q_o, where q is the electron production rate at various moments of solar occultation and q_o is the electron production rate before the eclipse.

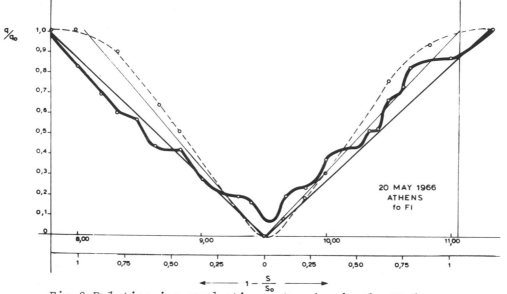

Fig.8. *Relative ion production rate q/q_o in the F1 layer, versus optical occultation(dotted curve),occultation of only the solar disc $I-S/S_o$ (light line)and occultation of the solar disc and the corona.(full line)*

Fig.8 shows the relative production rate q/q_o compared with optical occultation during the eclipse of the F1 layer.Deviations of the ionization rate are not consistant with the contribution from the eclipsed Sun during the occultation and the uncovering eclipse period. The decrease of the ionization during the covering of the West active part is also followed by a similar decrease during its uncovering which is not consistant with the distribution of the radiation already discussed. The same remark is also applied by comparing q/q_o with the variations of $I-S/S_o$ (S_o=surface of the solar disc, S = its covered part).This result could be expected because the above curves refer to the occultation of the solar disc without including its corona. By considering the occultation of both the corona and the solar disc we obtain the best results justifying the assumption made.The same results are obtained by applying the above method to the foE values (fig.5) measured by the Constantinople ionosonde (private communication).Experimental data are well fitted only when we consider the ionization produced by both the corona and the solar disc. Fig.9 shows the increased

importance of the west limb of the corona producing a decreased
ionization on the E layer during its occultation.

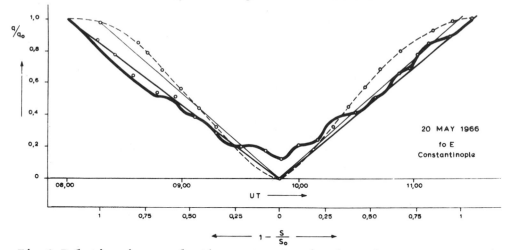

Fig.9 *Relative ion production rate* q/q_0 *in the E layer,versus optical occultation (dotted curve),occultation of only the solar disc*
$I-S/S_0$ *(light line) and occultation of the solar disc and the
corona (full curve)*

From the present analysis of the F1 and E-layer variations we
may conclude that an important part of the total ionizing radiation
of the Sun is emitted from a bright corona. Active centers distributed in the surface of the solar disc are of little contribution
to the ionization of the F and E layer by their soft x-rays. In
the contrary hard x-rays produced by the above active centers are
responsible for the D-layer variations.

F2 – region

The annular solar eclipse over the North part of Africa and
the East part of Europe on 20 May, 1966, caused an increase of
foF2, similar to the one observed during previous eclipses and
especially the total solar eclipse over the North part of America
and Canada on 20 July 1963.

Evans (1965) in discussing this particular behavior observed
during the solar eclipse of 20 July 1963 and on the basis of radar
backscatter measurements, suggested that increases may be related
to an increase of Nmax. during the first half of the eclipse, due
to a rapid downward diffusion of ionization from above hmaxF2. In
order to induce a rapid downward diffusion of ionization simply as
a result of temperature effects, it is necessary that the exospheric
electron temperature Te should be much larger that the neutral temperature Tn.

Hanson (1963),Evans (1964) and Bourdeau (1964) suggested that the extent to which Te exceeds Tn is a marked function of latitude, and that we might not expect eclipses that occur at low latitudes to cause very large decreases of Te. Another consideration is that the magnetic dip is controlling the rapidity with which the upper F-region can responded to given changes of temperature. Following the above theories an increase in foF2 must depend to a great extend on the dip, which according to Evans (1965) must be at least of the order of 60^0.

Fig 10 shows the variation of foF2 observed by all stations listed in Table II. We can see that at Dakar, with the maximum of obscuration a large decrease of foF2 was observed, throughout the whole eclipse period. The only peculiarity is that the minimum of foF2 presented a slight delay after the maximum of obscuration which occurred at 08.08 LT. At Ouagadougou, where the percentage of obscuration was 89 per cent, the eclipse produced first an

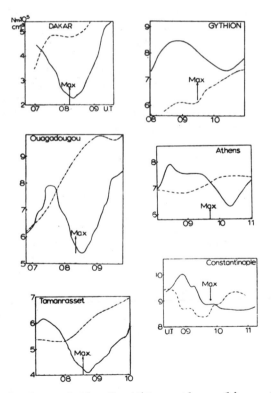

Fig.10.The variation of the NmaxF2 on the eclipse day(solid line) and the previous control days(chain lines)at Dakar,Ouagadou-gou,Tamanrasset,Gytheion,Athens and Constantinople

increase, followed by a large decrease, At Tamanrasse, located on
the ground center eclipse line, where the percentage of obscuration
was 87 per cent, the foF2 increase was observed before the beginning
of the eclipse and it was followed by a smaller decrease than that
at Ouagadougou. The eclipse maximum at Tamanrassed occured at
08.34. The eclipse path then crossed the Mediterranean sea, and 1
hr later, Gytheion ionosonde in the South of the Peloponnese, observed an increase, from the beginning till the end of the eclipse. It
is of importance to note that the Gytheion control curve is plotted
from data before and after the eclipse day.

In considering the variability of foF2 values, there seems
little to be gained by obtaining a mean of many other days for
comparison..

Athens (Scaramanga) with a maximum of obscuration 98.8 per cent
at ground level, presented an increase of foF2 during the first
half of the eclipse period, of the order of 0.75 Mc/s and a decrease
of the order of 0.8 Mc/s for the second half, from eclipse maximum
(09.31) to the end. The same behavior was observed by Istanbul ionosonde, with the only difference of a slight delay at the beginning
of the decrease.

T A B L E   III

Year of eclipse	Latitude	Longitude	Mag.dip	Time max.obscur.	Effect on F2
1940 (PIERCE)	$31°$S	$26°$E	64	16.00	Increase followed by decrease.
1944 (LEDIG)	$12°$S	$75°$W	0	09.00	-"-
1952 (LEJAY)	$4°$N	$19°$E	12	08.21	New layer formed
1963 (EVANS)	$61.2°$N	$149.9°$W	74	10.00	Increase
1966 (ANASTASSIADES)	$38°$N	$24°$E	53	11.31	Increase followed by decrease.

All above observations indicate, that during the solar eclipse
of 20 May 1966, an increase of foF2 followed by a decrease, was
produced.

A decrease was observed only by ionosondes located at places
where the percentage of obscuration was less than 80 per cent.

The beginning decrease was delayed as a function of increasing
the latitude and the magnetic dip.

The virtual heights along the path were observed to be different. In general, a lowering of hmaxF2 was observed.

An increase of foF2 may also be deduced from satellite observations, made during this same solar eclipse.(Anastassiades and Matsoukas) During a part of the first half, S-66 satellite crossed the
eclipse path from North to South. Fig. 11 shows the electron content,
measured during the eclipse day at different subionospheric latitudes,
compared with electron content measured from revolutions of the same
satellite, crossing the same ionospheric area at previous control
days and essentially at the same hours. In assuming that
the maximum contribution to the electron content is due to the F-
region, the increase of electron content observed during the eclipse
day may be oonsidered as consistent with the foF2 increase. The
difference in slopes between control day curves, is the subject of
another paper now under preparation.

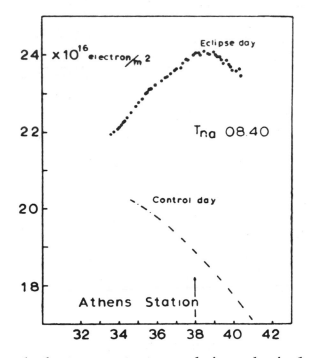

Fig.11 *Total electron content vs.sub-ionospheric latitudes,*
measured from S-66 active satellite amplitude recor-
dings of revolution 8068 North-South at 08.40UT,
20 May 1966

EARLIER ECLIPSE OBSERVATIONS

Observations of eclipse effects on the F-region, are now available from almost three complete solar cycles. Several eclipses produced an increase of foF2, but solar eclipse of the 20 May 1966 is the only one to present this behavior of foF2, for the path traversing the North African and South European areas.

Table III gives the list of all previous eclipses, during which an increase of foF2 was reported.

All eclipses listed above occurred near or in years of sunspot minimum. Their percentage of obscuration was over 80 per cent. The appearance however of an increase in foF2 is not dependent upon latitude or magnetic dip. The eclipse reported by Ledig (1946),who made observations in Huancayo (12°S,75W) at the magnetic equator during 1944, year of sunspot minimum, presented an increase followed by decrease, in spite of the fact that the latitude was low and the magnetic dip was 0 (Fig. 12). On the contrary during the next sunspot minimum, in 1954, only a decrease is reported by Stoffergen,despite the fact that latitude and dip were high (60N,18E,dip 75).

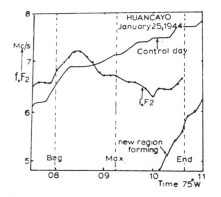

*Fig.12 Variations of foF2 on the eclipse day
(25 January 1944) and control days,at
Huancayo.*

Fig.13 shows times of occurrence of eclipses with an increase in foF2 during the period of three complete solar cycles. We can see that all of them occurred when the Wolf number was less than 60. The behavior of foF2, in 1954, year of sunspot minimum, is difficult to interpret. Control day curves are missing from observations reported by Stoffergen. A similar comment may be made for curves published by Dominici, from the same eclipse, the control day curve being plotted only half an hour after the beginning of

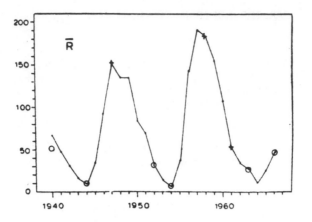

Fig.13 Solar eclipses with an increase
       (circles) of foF2 during the period
       of three somplete solar cycles.Eclipses
       with only decrease are indicated by
       a cross.

the eclipse. More reliable are the results reported by Davies from
observations made during the same eclipse at several places in
Canada. The most important for our discussion are the control and
eclipse day curves, obtained in Ottawa, where the percentage of
obscuration at a height of 200 km was 90 per cent (Fig.14).Half an
hour after the eclipse started, control and eclipse day values of
foF2 were the same. No increase is reported, but the delay of the
decrease start is typical.

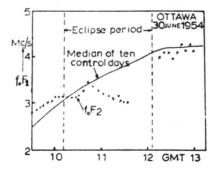

Fig.14 Variations of foF2 on the eclipse day
       (30 June 1954) and control days, at
       Ottawa.

Of special importance are the results reported by Shimazaki following his measurements in Rio Grande, Brazil during the 12th November 1966 solar eclipse, that means six months after the eclipse over Europe. The Brazil eclipse presents several similarities with the European one. Both occurred during the local spring at midday and with a high occultation of the Sun. Fig. 15 shows the temporal variations observed in the electron density at various heights on the eclipse and control day. Variations in the maximum electron density NmaxF2 are shown by thick curves.The three long lines on the absciss indicate the first contact, the totality and the last contact on the ground respectively.

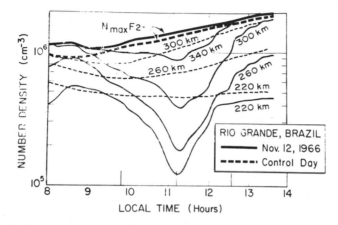

Fig.15 Variations in the electron density at various heights on the eclipse and control day et Nov. 12,1966.

An increase is observed in the number density but before the eclipse began. The behavior is similar to that of Tamanrasset on May 20, 1966 in Algeria. Both locations are symmetrical to the magnetic equator, being 30° dip angle North and South. Maximum electron density in Rio Grande is increased (Fig.14) before the beginning and for the entire eclipse period. The shape is not very pronounced as in Gytheion, however it is evidently an increase not followed by a decrease.

Thomas and Robbins (1956) have tried to evaluate the transport term using the production and loss rates for electron tentatively suggested by Ratcliffe (1956). They obtained the same temporal variation of transport for the eclipses in Cambridge (1954) and Chartoum (1952)).

Shimazaki calculated the vertical velocity drift in the F region during the eclipse over South America. He found that an additional velocity is largely upward over Rio Grande which amplifies the velocity drift on the control day.This amplification of the vertical velocity drift in the F region during the eclipse may be explained according to Shimazaki by the effect of an additional polarization electric field developed in the area of reduced conductivity in the lower dynamo region.The additional velocity drift might instead become slightly downward if the F region were connected with some area adjacent to the region of reduced conductivity.

CONCLUSIONS

Critical frequencies variations in the E and F1 layers,during the May 20 1966 annular solar eclipse,can not be attributed to a uniform solar disc as the only ionizing agent.An important contribution from a bright corona close to the solar limb must be considered.In a first approximation,the contribution of each of the above mentioned ionizing agent,based on measurements on 2980 MHz, indicate that 70% is due to the uniform solar disc and 30% to the bright corona close to the limb.This amount of 70% must be reduced even more in order to take in account the contribution of the three active sources distributed in the surface of the solar disc.If,following our absorption measurements on 1,98 MHz made in Athens and absorption measurements on 3,85 MHz made by Bischoff and Taubenheim in Mitchurin,we consider that active sources emitting hard X-rays produced the ionization in the D-region,we may assume that their contribution combined with the ionizing radiation of soft X-rays from the corona close to the limb,did not exceed 5%.Similar to the above percentages of contribution are indicated by Taubenheim and Serafimov in their recently published analysis of the E-region during the May 20,1966 annular solar eclipse.

foF2 increases were observed at five widely separated stations (Quagadougou,Tamanrasset,Gytheion,Athens,Istanbul)during the 20 May 1966 annular solar eclipse.At stations,more distant from the ground eclipse path,the effect on foF2 was less pronounced or even absent.In all stations from which increase in foF2 was observed the percentage of obscuration was higher than 85 per cent at F-region height.

An examination of previous eclipses shows that increases appears mainly at years near sunspot minimum.This is the case for 1940,1944 1952,1963 and 1966 solar eclipses.Experimental observations reported for the 1954 solar eclipse,year of sunspot minimum must be considered with circumspection.Following DALGARNO(1963) and HANSON (1963) the decreased electron density N ,during years near sunspot minimum,should cause the temperature difference Te-Tn to increase. BOURDEAU (1964) suggestion that Te > Tn, above hmaxF2 only during years of sunspot minimum,is consistent with the behavior of the F-region,observed during the 20 May 1966 and the above mentioned earlier solar eclipses.

EVANS (1965) and BOURDEAU (1964) have suggested that the extent to which Te exceeds Tn is a marked function of latitude. EVANS (1965) indicated that there seem to be two necessary, but possibly not sufficient, conditions for observing an increase in foF2 during solar eclipse. First, the eclipse should be total or very nearly so at F2-region heights, and secondly that the magnetic dip should be large, higher than 60°.The first condition,concerning the eclipse totality was well fullfilled by all eclipses which occurred near years of sunspot minimum.Latitude however and magnetic dip, seems to control, to a less extent,the appearance of an increase in foF2.African and European observations during the solar eclipse of 20 May 1966,show that latitude and dip as they become higher,have a definity influence,only on the progressing significance of the foF2 increase.Ouagadougou,Tamanrasset and observations from other low latitude stations, show that an increase in foF2,may be expected during eclipses that occur at years near sunspot minimum,also in low latitudes and for magnetic dip much less than 60°.

REFERENCES

ANASTASSIADES and MORAITIS	1968	*J.Atm.Terr.Phys.V.30 p.1471*
ANASTASSIADES and BOVIATSOS	1968	*Nature V.219 pp 1139-1141*
BISHOFF and TAUBENHEIM	1967	*J.Atm.Terr.Phys.29,1062*
BOUSQUET C.,LAUFEUILLE M., VASSEUR G.et VILA P.	1967	*Annls Geophys. 23,345*
BOURDEAU R.E.	1964	*COSPAR*
DALGARNO A.M.	1963	*Planet,Space Sci 11,463*
DAVIES K.	1956	*Solar Eclipses and the Ionosphere Pergamon Press,Oxford*
DOMINICI P.	1956	*Solar Eclipses and the Ionosphere Pergamon Press Oxford*
EVANS J.V.	1965	*J.Geophys.Res. 70,3*
HANSON W.B.	1963	*Space Research III 283*
LEDIG P.G.	1964	*J.Geophys.Res. 51,411*
LEJAY R.P.	1956	*Solar Eclipses and the Ionosphere Pergamon Press Oxford*
PIERCE J.A.	1948	*Proc.IRE,36,8.*
STOFFENGEN W.	1956	*Solar Eclipses and the Ionosphere Pergamon Press, Oxford*
THOMAS and ROBBINS	1956	*Solar Eclipses and the Ionosphere Pergamon Press,Oxford*
TAUBENHEIM and K.SERAFIMOV	1969	*J.Atm.Terr.Phys.Vol.31,pp.307 to 312.*

L'ECLIPSE DE SOLEIL DU 20 MAI 1966

I.Ozdogan - T.Bulat

Universite d Instanbul

I . INTRODUCTİON
    Des observations ionosphériques de l'éclipse circulaire de
20 Mai 1966 ont été effectuées à İstanbul.La situation géographique de la station ionosphérique est de:
                41° 08' N ; 29° 02' E .
    La ligne centrale de l'éclipse traverse la Méditerranée,
la Grèce du sud vers le nord et la Turquie dans la direction
SW (Fig.1)et la trace de la trajectoire de la lune sur le disque
solaire est comme indique la figure 2.

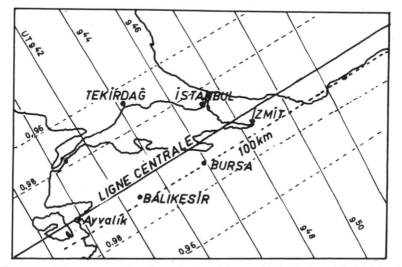

Fig.1 :Ligne centrale et la ligne à 100 km.d'altitude·

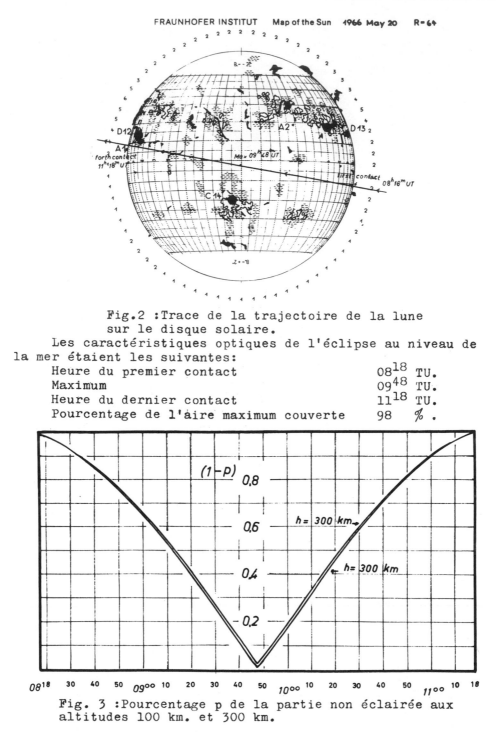

Fig.2 :Trace de la trajectoire de la lune
       sur le disque solaire.

Les caractéristiques optiques de l'éclipse au niveau de
la mer étaient les suivantes:

Heure du premier contact 08^{18} TU.
Maximum 09^{48} TU.
Heure du dernier contact 11^{18} TU.
Pourcentage de l'aire maximum couverte   98   % .

Fig. 3 :Pourcentage p de la partie non éclairée aux
altitudes 100 km. et 300 km.

La figure 3 représente,aux altitudes 100 km. et 300 km.le pourcentage p de la partie occultée aux différents moments de l'éclipse.

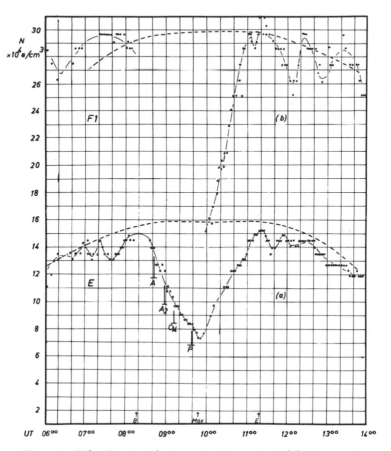

Fig.4 :Résultat d'observation de l'éclipse de 20 Mai 1966 et la courbe médiane de cinq jours autour du jour d'éclipse.

               a.Couche E ;   b.Couche F_1 .

Le sondeur utilisé est du type panoramique et enrégistre d'une façon continue.La puissance à la crête est de 20 KW.Le sondeur émet d'une façon continue des fréquences depuis 1 MHz jusqu'à 20 MHz .La durée d'une pulse est de 100 micro seconde et le nombre des pulses par seconde est de 66.

La précision obtenue à partir des ionogrammes sont:

± 2 km. pour la hauteur virtuelle h' des couches E et F_1 et de ± 5 km. pour F_2 .

± 0,025 MHz pour les fréquences critiques des couches E et F_1 , ± 0,1 MHz pour celle de F_2 .

II . RESULTATS

Nous avons voulu montrer avec les figures 4 , 5 et 6 les résultats d'observations concernant les différentes couches.

1 .Sur la figure 4a ,nous voyons la distribution de la densité électronique maximum N pendant les heures d'éclipse,en même temps que sa valeur médiane correspondant aux cinq jours autour du jour d'éclipse.Nous ajoutons que pendant ces dix jours de controles,les mesures sont effectuées toutes les dix minuttes tandis que pour le jour d'éclipse elles sont faites à l'inter-valle de trois minutes.

Nous voulons formuler certains observations concernant la figure 4a,à savoir :

i .L'effet d'occultation de chacun des centres actifs,mar-qués sur la surface solaire par les caractères A , A_2 et C_{14} (Fig.2),est bien distinct.Sur la figure,des décroissement dans les valeurs de N sont marquées aux heures 8^{42} , 8^{58} et 9^{12} ,qui correspondent-croyons nous bien- respectivement aux

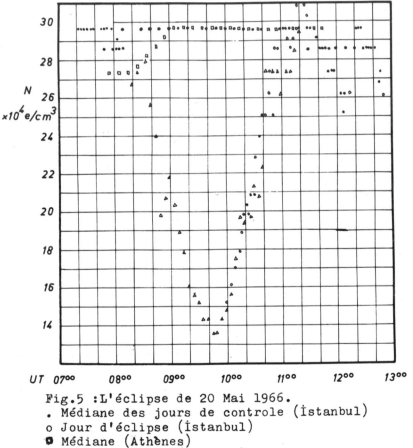

Fig.5 :L'éclipse de 20 Mai 1966.
. Médiane des jours de controle (İstanbul)
o Jour d'éclipse (İstanbul)
◙ Médiane (Athènes)
△ Jour d'éclipse (Athènes)

centres cités plus haut.

    ii .L'accroissement bien net dans la valeur de N ,vue de
même,sur la figure juste à 9 ^{38} TU a eu lieu,à l'erreur expé-
rimentale près,à l'instant que le bord NW du disque solaire sort
de l'ombre.Il nous semble que cet accroissement est dû, de même,
à l'effet des protubérances qui y sont présentes.

    iii .Les oscillations dans la valeur de N que nous voyons
avant et après l'éclipse sont également attribuées à l'effet
des bords du disque solaire.

    2 . Les valeurs d'observations concernant la couche F_1 sont
portées sur la figure 4b. Ces valeurs sont obtenues de la même
façon que précédement expliquée pour le cas de la figure 4a.
La courbe d'éclipse pour F_1 est incomplète,une couche sporadique
E intense occultait dans l'intervalle entre 08 ^{11} - 10 ^{00} TU la
lecture de f_oF_1 .

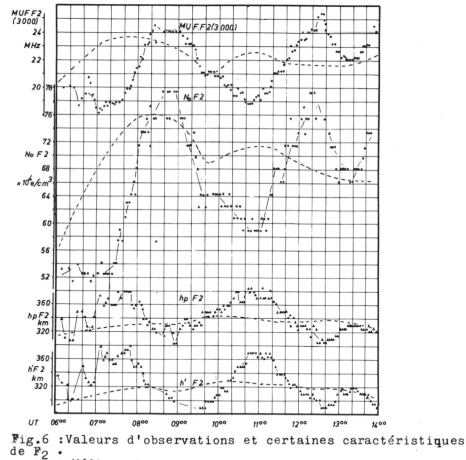

Fig.6 :Valeurs d'observations et certaines caractéristiques
de F_2 .
---- : Médiane des jours de controle.
-o-o-): Jour d'éclipse.
-Δ-Δ-)

Nous avons voulu compléter l'intervalle occulté en utilisant les résultats d'Athènes de 286 km. dans la direction SW (Latitude géographique 38° 01' N.). La figure 5 représente l'ensemble des déterminations de deux stations, İstanbul et Athènes.

3 . Sur la figure 6 sont présentées les valeurs d'observation et les caractéristiques de la couche F_2, obtenues, toujours, suivant le même procédé utilisé dans les cas précédents. Nous y voyons:

MUF (3000) F_2 ; N_0F_2 ; h_pF_2 et $h'F_2$.

Comme on s'y attendait, l'effet de l'éclipse est nettement marqué sur ces grandeurs.

III . DISCUSSİON

Pour l'analyse des observations effectuées pendant l'éclipse solaire, nous admettons que la surface optique du soleil émet d'une façon continue et uniforme des radiations ionisantes, sur laquelle se superposent des zones plus ou moins actives. Si on omet le terme de diffusion on a par suite l'équation d'équilibre suivante:

$$\frac{dN}{dt} = (q_0 + q_1) \cos \chi - \alpha N^2 \qquad (1)$$

où sont:

N :densité maximum des électrons

α :coefficient de recombination

$q_0;q_1$:ionisations dues respectivement à la totalité du disque solaire et aux sources localisées ne suivant pas la même loi d'occultation

χ :distance zénitale de soleil .

Pour le cas d'éclipse, le terme q_0 sera multiplié par un facteur $(1-p)$ où p est la fraction occultée du disque solaire qui varie avec le temps t.

Soient N_0 et N_1 les valeurs de N correspondant, respectivement, à la courbe médiane et à celle de jour d'éclipse (Fig.4a et 4b). On obtient $N_0 = 15,8.10^4$ e/cm³ et $N_1 = 7,4.10^4$ e/cm³ . A l'instant de maximum occultation, les dérivées de N sont toutes nulles. Nous tirons:

$$(N_1 / N_0)^2 = \frac{q_0 + q_1 (1-p)}{q_1 + q_0} = 0,22 ,$$

ou encore:

$q_1 / q_0 = 0,25$,rapport qui est relativement grand pour être expliqué avec l'effet des sources isolées sur le disque. Il est suggéré alors[x] qu'un effet des bords positifs appréciable est venu s'ajouter à ces centres se traduisant par une augmentation de l'ionisation. En admettant une surface effective S_{eff} plus large que celle optique S_{op} ,nous avons voulu calculé, avec diverses valeurs de α, la courbe théorique .Nous avons prie pour cela $S_{op}/S_{eff} = 80/100$ .Le pourcentage de la surface occultée pendant l'éclipse sera donc P = 1,25 p .

[x] Voir Bowhill S.A., et al 1969.

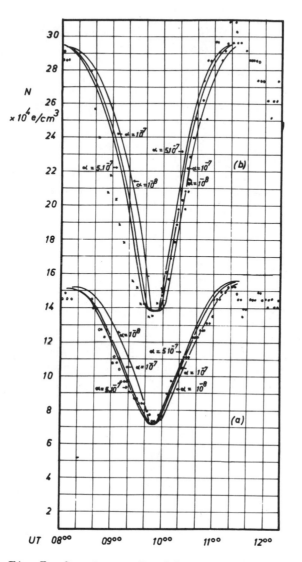

Fig.7 :Courbes calculées pour le jour
d'éclipse. a. Couche E ;   b.Couche F_1,
complétée avec les valeurs d'Athènes .

   Une solution générale de l'équation d'équilibre est donnée
par Rydbeck et Wilhelmsson (1954).Nous avons utilisé l'équation
4 de ces auteurs où nous avons mis,à la place de q,la grandeur
$Q = q_0 (1-P)$ et nous y avons négligé le terme Q''/Q devant l'uni-
té.Il vient en suite:

$$N = (Q/\alpha)^{-1/2} \left(1 - \frac{Q'}{4Q\sqrt{\alpha Q}} - \frac{Q'^2}{8\alpha Q^3}\right)$$

où, $Q' = dQ/dt$.

1 . Eclipse dans la couche E.

Nous avons reproduit sur la figure 7a le résultat de calculs effectués avec les valeurs suivantes de α:

$$\alpha = 5 \cdot 10^{-7} ; \quad = 10^{-7} ; \quad = 10^{-8}. \tag{3}$$

Nous remarquons que la courbe de Rydbeck correspondant à la valeur $\alpha = 5 \cdot 10^{-7}$ s'adapte mieux à l'état d'ionisation concernant la couche E.

2.. Eclipse dans la couche F_1.

Les courbes théoriques sont calculées ici en utilisant les valeurs de α de la suite (3). L'examène des courbes (Fig. 7b) nous permet de conclure que la valeur du coefficient de recombinaison valable pour la couche F_1 se situe entre les valeurs de 10^{-7} et de 10^{-8}.

LITERATURE

Bible, K., 1960, Ann. Géophysique, T. 16, No. 1.

Bible, K., et Delebeau, F., 1955, Z. Geophysik.

Delebeau, F., 1953, Ann. Géophysique, T. 9, No. 4.

Bowhill, S. A., et al., 1969, Solar Eclipse and Ionosphere.

Rydbeck, O. E. H., and Wilhelmsson, H., 1954, Trans. Chamers
     Univer. Tech. No. 149.

QUELQUES RESULTATS PRELIMINAIRES D'OBSERVATIONS IONOSPHERIQUES

LORS DE L'ECLIPSE DU 20 MAI 1966

A.Haubert

Groupe de Recherches Ionospheriques France

I - RESULTATS DES MESURES DE TAMANRASSET ET OUAGADOUGOU

Quoique n'ayant pas lui-même participé à l'étude de cette éclipse, le conférencier présente les résultats obtenus par ses collègues du GROUPE DE RECHERCHES IONOSPHERIQUES (G.R.I.) et les fait suivre de quelques commentaires personnels.

Les résultats présentés sont extraits des publications sui-vantes : Note Technique du G.R.I. n° 66, 1966 et Annales de Géophy-sique, T 23, n° 3, 1967.

Cette étude a porté sur les mesures ionosphériques effectuées en six stations africaines, à savoir : TAMANRASSET, OUAGADOUGOU, DAKAR, IBADAN, BANGUI et PORT-GENTIL. Les résultats saillants peuvent être résumés comme suit, pour chaque région ionosphérique.

Région E. Pour l'ensemble des stations, une couche Es occultante a empêché l'étude de la région E normale, cependant, à TAMANRASSET et à OUAGADOUGOU, des interruptions momentanées de l'occultation ont permis, avec quelques interpolations de traiter les ionogrammes par la méthode de Doupnik et Schmerling et d'obtenir des tracés conve-nables des hauteurs vraies pour la région F.

Région F1. De la comparaison de la courbe de N_mF_1 à OUAGADOUGOU et des enregistrements de flux solaire sur 9,1 cm, 21 cm, et 26-40 A, il est apparu que la meilleure corrélation était fournie par le rayonnement sur 9,1 cm.

Une étude quantitative plus poussée a fait ressortir certaines anomalies pour l'explication desquelles diverses hypothèses sont proposées par les auteurs, en particulier la possibilité d'une variation du coefficient de recombinaison α.

Région F2. Un premier examen des mesures de TAMANRASSET a permis de déterminer, à l'altitude de 280 km, les valeurs suivantes pour les paramètres classiques :

$\beta \simeq 3.10^{-4} \ S^{-1}$
$q6 \simeq 2$ à $3.10^2 \ cm^{-3} \ S^{-1}$

Une autre méthode de détermination de β, basée sur le retard entre le minimum de densité électronique à une altitude donnée et le maximum de l'éclipse astronomique, a conduit à formuler (pour TAMANRASSET) une loi approchée valable entre 240 et 280 km :

$$\beta = 1,6 \ . \ 10^{-4} \ exp \left[\frac{280 - h}{28} \right] \ S^{-1}$$

En dessous de 240 km, la loi de variation de l'ionisation en fonction du temps cesse d'être représentée par une équation différentielle du premier ordre et β ne peut plus être considéré comme une constante. Au-dessus de 280 km les auteurs attribuent à la diffusion le désaccord des mesures avec la courbe théorique.

II - COMPARAISON DE CES MESURES AVEC CELLES DES AUTRES STATIONS

Port-Gentil. L'intérêt de cette station est d'être magnétiquement conjuguée de TAMANRASSET. Il a été constaté que la variation de h_mF_2 était décalée vers l'instant de l'éclipse au point conjugué et que celle de N_mF_2 semblait aussi significative dans ce sens.

Dakar. Les auteurs remarquent un retard du minimum de N_mF_2 par rapport à la variation du rayonnement solaire.

Bangui et Ibadan. Aucune conclusion n'a pu être tirée des enregistrements par suite de leur qualité insuffisante.

III - COMMENTAIRES SUR CERTAINS POINTS PARTICULIERS

Après avoir résumé les conclusions des auteurs, le conférencier énonce quelques remarques personnelles quant aux points suivants.

a) Phénomène d'Eclipse Conjuguée. Ce phénomène a été signalé dans une Note du J.A.T.P., vol. 25, pp 105-107, à propos de l'éclipse du 11 Août 1961 et surtout de celle du 15 Février 1961. (J.A.T.P. de 1963).

Une étude récente faite par P. VILA du G.R.I. a montré que
l'ionisation maximale de F_2 était fortement diminuée sur le
bord sud de la coquille magnétique d'Apex 700 à 800 km,
conjugué tropical de la zone d'éclipse maximale à 08 h 45 TU
le 20 mai 1966.

Il semble donc qu'il faille porter une attention particu-
lière aux zones conjuguées des futures éclipses.

b) Phénomène du Retard des Minimums d'Ionisation. Le conférencier attire l'attention sur l'intérêt de cette étude. Il
rappelle une note qu'il a publiée au J.A.T.P. (vol. 24, pp
661 - 663, 1965) sur l'interprétation des constantes de
temps de l'ionosphère.

Il observe que si l'on fait état de la variation de β en
fonction de l'altitude, les retards des minimums de densi-
tés devraient croître de plus en plus rapidement avec
l'altitude, alors que les retards observés, s'ils croissent
en effet assez rapidement à partir de 200 km, tendent
ensuite vers une limite de l'ordre de vingt minutes.

Ce phénomène n'est pas constaté uniquement lors d'une éclipse, mais aussi au lever du soleil, comme le montrent, par
exemple, les courbes tracées pour les jours de contrôle de
l'éclipse du 15 février 1961.

Tout semble se passer comme si le maximum de production
photoïonique était situé à la base de la région F ; les régions situées au-dessus n'étant photoïonisées qu'à un moin-
dre degré, l'ionisation constatée étant complétée par
diffusion des régions inférieures d'une part et de l'exos-
phère d'autre part (l'ionisation de la région F au lever du
soleil - Proc. Nato Adv. Studies Inst., Skeikampen, April
1963).

c) Variation de α dans la Région F. Au cours de l'étude de
l'éclipse du 15 février 1961 à GARCHY, le conférencier était
arrivé à la conclusion que l'hypothèse de la constance de α
était sujette à caution, même les jours sans éclipse et
surtout au lever du soleil.

En ce qui concerne l'éclipse du 5 février, il avait formulé
l'alternative suivante : ou bien l'on considère α constant
et il faut élargir le soleil ionisant dans des proportions
considérables, ou bien l'on considère que le rayonnement
ionisant reste proportionnel à la surface non éclipsée du
soleil, alors il faut admettre que α varie en synchronisme
avec la variation du rayonnement.

L'étude ionosphérique des éclipses n'est donc nullement épuisée mais, comme l'a dit si bien, le Professeur BOWHILL, à l'ère des satellites artificiels tous les espoirs sont permis. L'on dispose en effet maintenant des sondages par en haut complétant les sondages par en bas et aussi, et surtout, de la possibilité d'observer, hors de l'écran atmosphérique, les divers rayonnements solaires.

CHANGES IN THE TOPSIDE IONOSPHERE DURING SOLAR ECLIPSES

P. A. Smith and J. W. King

S.R.C.,Radio and Space Research Station, Ditton Park,

Slough, Buckinghamshire, England

ABSTRACT

Topside sounder data, obtained by means of the Alouette 1 and Alouette 2 satellites, are used to determine the behaviour of the electron concentration in the upper F-region during three solar eclipses in the South Atlantic area. The concentration decreased at all heights and locations studied, but the decreases observed were greater near 400 km than below this height. It is apparent that the observed behaviour may be explained by a downward movement of ionisation caused by a decrease in plasma temperature as well as a decrease in the ionisation production rate.

INTRODUCTION

Only a limited number of observations of the electron concentration at different heights in the upper F-region have been made during solar eclipses. Evans (1) reported the changes in electron concentration, observed using the incoherent backscatter technique, which took place during the 20 July 1963 eclipse and King et al. (2) used topside sounder data to study the behaviour during the eclipses of 25 January 1963 and 14 January 1964. In the following section Alouette 1 data, acquired during the two latter eclipses, are compared with available control information in order to determine the percentage by which the electron concentration changed as a result of the eclipses; data obtained by means of the Alouette 2 satellite during the eclipse of 12 November 1966 are also treated in the same way. The actual percentage changes have been calculated for all heights between the satellite and the F2-layer peak at different locations in the South Atlantic region during the three eclipses. The data available are such that it is not possible to follow the changes at any one location throughout the eclipse, but the interesting differences between the observed behaviour at different heights and latitudes can be investigated.

CHANGES OBSERVED DURING THE ECLIPSES

25 January 1963

Measurements were made, by means of the Alouette 1 satellite, of the vertical distribution of electron concentration at a number of latitudes in the eclipsed region. The distributions have been compared with data obtained at corresponding latitudes during a pass just prior to the start of the eclipse and also two passes on the following day. The results obtained near 60°S magnetic dip are shown in Table I; the eclipse values are much lower than those obtained during any of the three control passes. The local times at which the various passes occurred are very nearly the same and, therefore, the results would not be expected to be affected by local time differences. Although the data were acquired at different longitudes, it is apparent from Table I that longitude differences of the order of 25° (between the consecutive passes on 26 January) do not introduce serious complications. It seems reasonable, therefore, to assume that all the non-eclipse measurements can be used as control data and that the differences between the various control data indicate the magnitude of the errors inherent in this type of analysis.

The percentage reductions in the electron concentration have been calculated for different magnetic latitudes and some typical results are shown in Fig. 1. The reduction produced by the eclipse exceeds 25% in all cases. The greatest reductions tend to occur near 400 km and this feature, which was also observed in the two other eclipses studied, will be discussed in a later section.

Table 1. Details of measurements made near $60°$S magnetic dip in
order to study the annular eclipse of 25 January 1963

Height (km)	Electron concentration ($\times 10^3 cm^{-3}$)			
	Eclipse pass	Control data		
1000	11	18	23	18
900	14	24	30	24
800	18	33	40	34
700	24	47	56	48
600	33	72	84	74
500	51	125	136	128
400	89	256		266
300	182	497		
Date	25 Jan	25 Jan	26 Jan	26 Jan
U.T.	1314	1130	1021	1207
Longitude	$71.6°$W	$42.2°$W	$29.6°$W	$53.2°$W
Latitude	$65.5°$S	$68.2°$S	$65.8°$S	$68.3°$S
Duration of eclipse (minutes)	118			
Time after middle of eclipse (minutes)	34			
L.M.T.	0828	0841	0823	0834
Magnetic dip at ground level	$60°$S	$60.5°$S	$60°$S	$60.5°$S

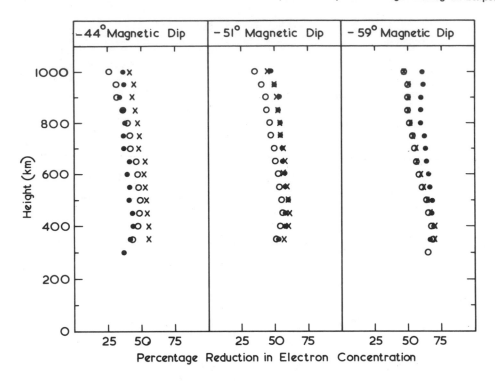

Fig. 1. Percentages by which the electron concentration
was reduced at different latitudes during the 25 January
1963 eclipse. In each case, the electron distribution
observed during the eclipse has been compared with three
control distributions; the open circles represent
comparisons with a pre-eclipse pass on the same day and
the solid circles and the crosses indicate comparisons
with passes on the following day.

14 January 1964

Measurements of electron concentration were made on the
eclipse day during three consecutive passes of the Alouette 1
satellite. The times of these passes were such that data were
obtained shortly before maximum solar obscuration and at two
later times, both of which were after the eclipse. King et al.
(2) have shown how the concentration at 400 km varied with
magnetic dip during these passes and they drew attention to the
observed gradual recovery from the low values which occurred
during the eclipse. In the present paper, control data acquired

one, three and five days before the eclipse (in the same geographic area and at similar local times) have been incorporated. The results for $60°$S magnetic dip are given in Table II and the observed percentage decreases are shown in Fig. 2.

Table II. Details of measurements made near $60°$S magnetic dip in order to study the partial eclipse of 14 January 1964

Height(km)	Electron concentration $(\times 10^3 cm^{-3})$					
	Eclipse data			Control data		
1000	11.7	16.1	19.0	20.3	17.3	20.0
900	15.0	20.7	23.8	25.6	23.2	26.9
800	19.0	26.9	30.6	33.5	32.8	38.4
700	25.2	36.2	41.7	45.8	49.4	58.1
600	36.4	53.4	61.7	68.7	80.8	101
500	60.0	88.8	103	115	150	201
400	114	162	200	221	319	423
300	235	293		410		
Date	14 Jan	14 Jan	15 Jan	13 Jan	12 Jan	9 Jan
U.T.	2105	2249	0036	2212	0029	2314
Longitude	$26.0°$W	$54.6°$W	$78.7°$W	$42.4°$W	$72.2°$W	$50.3°$W
Latitude	$61.5°$S	$64.1°$S	$60.8°$S	$61.5°$S	$62.4°$S	$63.5°$S
Duration of eclipse (minutes)	96	96	76			
Time after middle of eclipse (minutes)	−11	+93	+200			
L.M.T.	1921	1911	1921	1922	1940	1953
Magnetic dip at satellite	$60.3°$S	$60.3°$S	$59.8°$S	$59.8°$S	$60.1°$S	$60.1°$S

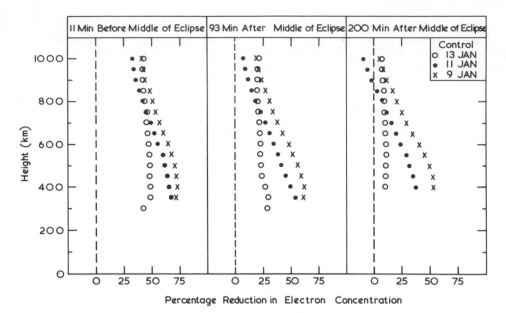

Fig. 2. Percentages by which the electron concentration
was reduced at 60°S magnetic dip during and after the
14 January 1964 eclipse. Each set of data obtained on
the eclipse day has been compared with the three sets
of control data; the results obtained using the
13 January control data (open circles) are likely to be
the most reliable because these control data were
obtained at local times closest to those of the eclipse
data.

    The differences between the results obtained using the
different sets of control data are quite significant; these
differences must arise partly because of day-to-day variability
and also, perhaps, because of differences between the local times
at which the various measurements were made. It is quite clear,
nevertheless, that the greatest decreases occur near 400 km as
reported above in the case of the 25 January 1963 eclipse, and
that the reductions are significantly smaller several hours after
the eclipse than near the middle of the eclipse. The behaviour
observed at other latitudes in the region studied was essentially
the same as that shown in Fig. 2. All these data indicate that
the ionosphere had returned to normal by about three hours after
the eclipse.

12 November 1966

Data obtained by means of the Alouette 2 satellite on the eclipse day were analysed in order to examine the electron distributions obtained at particular latitudes, but at three different longitudes corresponding to three satellite passes on the eclipse day. The results for 60° magnetic dip are shown in the right-hand section of Fig. 3; pass 4130 occurred 15 min before the start of the visual eclipse, while passes 4131 and 4133 were, respectively, 16 and 350 min after the end. The left-hand section of Fig. 3 shows two distributions obtained at a latitude which was just outside the eclipsed area on two orbits 123 min apart. There are only very slight differences between these two distributions whereas, inside the eclipsed area, a notable decrease occurred between passes 4130 and 4131; the subsequent return to the higher values observed during pass 4133 suggests that the low values were the result of the eclipse.

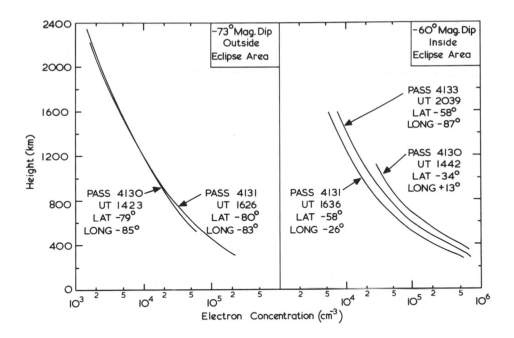

Fig.3. Vertical distributions of electron concentration obtained on 12 November 1966.

The different geographic latitude (34°S) at which the
magnetic dip was 60°S during pass 4130 means that it is
unsuitable as a control pass and, therefore, pass 4133 was used
to provide the control data (see Table III). The percentage
reduction has been computed and the results are shown in Fig. 4
together with data (for the same magnetic dip) from the other
eclipses.

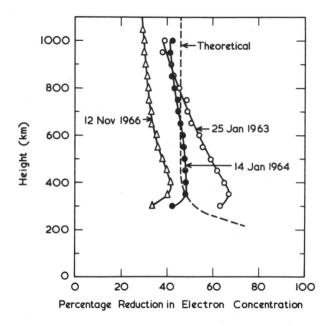

Fig.4. The percentages by which the electron
concentrations at 60°S magnetic dip were reduced as a
result of three different eclipses. The details of the
eclipses are given in the text. The portion of the
theoretical curve below 500 km shows the results of the
"fast diffusion" calculation made by Rishbeth (3) for a
hypothetical eclipse at middle latitudes; in order to
extrapolate this curve to greater heights it was assumed
that the percentage decrease above 500 km was the same
as that between 400 and 500 km.

Table III. Details of measurements made near 60°S magnetic dip in
order to study the total eclipse of 12 November 1966

Height (km)	Electron Concentration ($\times 10^3$ cm^{-3})	
	Eclipse data	Control data
1500	6.4	8.7
1400	7.6	10.4
1300	9.2	12.5
1200	11.1	15.4
1100	13.6	19.2
1000	17.1	24.5
900	22.2	32.1
800	29.6	43.6
700	41.8	62.8
600	64.4	99.8
500	107	174
400	199	343
300	440	667
Date	12 Nov	12 Nov
U.T.	1636	2039
Longitude	25.8°W	86.5°W
Latitude	58.3°S	57.8°S
Duration of eclipse (minutes)	134	64
Time after middle of eclipse (minutes)	83	405
L.M.T.	1453	1453
Magnetic dip at satellite	59.9°S	59.6°S

DISCUSSION

The behaviour of the upper F-region electron concentration, observed during the three eclipses studied, exhibits some common features. Decreases occur during and after the eclipses. Rishbeth (3) has shown that downward diffusion will result in decreases of electron concentration in the F-region; the results of his theoretical calculations for a hypothetical eclipse at middle latitudes (ignoring the effects of temperature and composition changes, but taking into account diffusion and the variations of production and loss) are shown in Fig. 4 by the dashed line. These predictions are in general agreement with the observed data except at the lowest heights.

A possible explanation of the behaviour observed below 400 km arises from the work of Evans (1) who found that, during the 20 July 1963 eclipse, decreases of electron concentration occurred at high altitudes whereas below 450 km the concentration increased at the beginning of the eclipse. He ascribed this increase to a redistribution of ionisation brought about by a downward movement caused by a decrease in the plasma temperature, and he suggested that the magnitude of the temperature decrease (and hence the decrease in the electron scale height) would determine whether or not any increase occurred in the concentration at the lower heights.

All of the days, except one, on which eclipse or control data were acquired were magnetically quiet; the exception is 9 January 1964 (one of the control days for the 14 January eclipse) which was moderately disturbed.

CONCLUSIONS

Changes in the electron concentration in the upper F-region during three eclipses have been studied by means of topside sounder satellite data. In all cases, the concentrations are found to be reduced during the eclipse and for several hours afterwards and, by using appropriate control data, the percentage decreases at different heights have been calculated. The greatest changes occurred near 400 km; below this height the decreases were not so marked. It seems likely that these changes, whether they resulted from movements of ionisation which replaced losses near the peak of the F-layer or from a temperature-induced contraction, indicate that ionisation diffused downwards throughout the topside ionosphere. In particular, the behaviour observed below 400 km appears to be at least partly the result of a temperature decrease.

ACKNOWLEDGEMENTS

The topside sounder satellites Alouette 1 and 2 were built by the Communications Research Centre, Ottawa and launched by the United States National Aeronautics and Space Administration. The work described in this paper was carried out at the Radio and Space Research Station and is published with the permission of the Director.

REFERENCES

1. J. V. Evans, J. Geophys. Res. 70, 131 (1965).

2. J. W. King, A. J. Legg and K. C. Reed, J. Atmos. Terr. Phys. 29, 1365 (1967).

3. H. Rishbeth, Proc. Phys. Soc. 81, 65 (1963).

CONCLUDING REMARKS

CONCLUDING REMARKS

Université de Paris, Physique de l' Atmosphère

Depuis déjà de nombreuse années, je suis chargé de la conclusion
de nos réunions. Aussi je ne puis me dérober à cet usage établi,
d'autant mieux que je dispose de l'assistance d'une personne plus
jeune et plu "dans le vent," dans le vent solaire devrait-on dire
ici.

Aussi je me bornerai à quelques généralités, et tout d'abord le
choix du sujet: "les éclipses de soleil et l'ionosphere."

Plusieurs fois, au cours des dernières années ce sujet a été
traité, notamment lors d'une réunion de la Commission Ionosphérique de l'A.G.A.R.D. Mais depuis, le développement des techniques
spatiales a permis de l'aborder sous un nouvel angle et de renou-
veler son intérêt.

Mais il y a aussi une autre raison, la Grece ayant été favorisée
par l'eclipse du 20 mai 1966, a de plus fait l'effort de recevoir
et de preter assistance à de nombreuses expériences étrangères.

Il était normal qu'une analyse des résultats obtenus permette de
se rendre compte que l'effort fourni alors n'était pas vain. C'est
ce qu'ont démontré les nombreux comptes rendus que nous avons
entendus.

C'était, d'autre part, pour ceux qui avaient participé à l'etude
de l'éclipse, le plaisir de se retrouver, dégagés des soucis d'une

---

†Professor E. Vassy died October 30, 1969. This is his last scientific
contribution.

expérience à ne pas rater (car on ne peut la recommencer le lende-
main) afin de mettre en lumière un progrès dans nos connaissances
et de se rémémorer dans le même cadre les palpitantes émotions
d'il y a trois ans.

Si les idées nouvelles nous arrivent sous l'influence de petits
chocs résultant de la confrontation de résultats expérimentaux et
d'efforts théoriques, intervient aussi le milieu dans lequel s'ac-
complit cet effort créateur.

Nos travaux ont mis en évidence la difficulté de notre problème.
Depuis que notre président HERBERT FRIEDMAN nous a montré l'existence du rayonnement X solaire, nos connaissances ont progressé à
grands pas, mettant en évidence leur complexité. Mais je n'entre-
rai pas dans plus de détail, laissant au Dr. BOWHILL le soin d'une
analyse plus détaillée.

Enfin je voudrais en votre nom féliciter et remercier l'artisan
de cette réunion le Pr. ANASTASSIADES pour sa brillante réussite.

Aussi devons-nous remercier l'Organisation du Traité de l'At-
lantique Nord d'avoir accepté de fincancer cette réunion. Elle en
a apprécié très certainment l'interêt scientifique et indépandam-
ment de la largeur de vues qui la caractirisé, elle sait quel
bénéfice tout progrès la connaissance de l'ionosphère entraine
dans la sécurité de ses télécommunications.

CONCLUDING STATEMENT

S. A. Bowhill

Aeronomy Laboratory

Department of Electrical Engineering

University of Illinois

Urbana, Illinois 61801

One of the most interesting and exciting aspects to emerge from this meeting is the excellent information beginning to become available from E- and Fl-region electron density measurements. Critical frequencies measured by ionospheric groups during the May 1966 eclipse are of very good quality, and the interpretation of the results will certainly be of great interest.

Several authors have pay attention to the question of the eclipse function for E- and Fl-region ionization. Soft x-rays in the 44-100 Å band certainly contribute to E-region ionization; perhaps equally important is the annular character of a solar eclipse in these wavelengths, because of the substantial contribution from the uneclipsed part of the corona. It seems certain that if a revised eclipse function is used, including a coronal contribution, the recombination coefficient for the ionization will be found correctly; however, if the optical disc eclipse function is used a recombination coefficient will be found which is substantially lower than the correct value.

I believe the evidence from x-ray measurements of the sun favors the revised eclipse function, and the high value for the recombination coefficient. During the March 1970 eclipse in the United States, some rocket experiments will be carried out where the intensity of x-rays will be measured, which will provide additional information concerning the existence of a substantial coronal contribution even at totality.

While the existance of individual sources of x-rays distributed on the surface of the solar disc has been demonstrated from measurements in D- and E-region ionization, the exact size of these sources is not known very definitely, and their intrinsic brightness is perhaps not known with great accuracy.

Assistance is needed from solar physicists as to the best index of brightness of these sources from radio noise measurements; to know which is the best combination to give an index of the ionizing intensity as a function of solar obscuration. It was also brought out at this meeting that x-ray intensity can vary even during the time of the eclipse.

301

The question of the brightness of the east and west limbs of the sun is very important, and obviously has a great effect on the interpretation of ionospheric measurements. Straka presented evidence that the minimum in radio emission is displaced in time from optical minimum, suggesting that time delays previously found in measurements of E-region ionization relative to totality may have been associated with a displacement in the minimum of solar ionizing intensity relative to the center of totality for the visible disc, rather than being due to a time delay because of finite recombination rate of the ionization.

In considering the role of two types of ions (for example, molecular oxygen and nitric oxide) in the E region, one must note the possibility that elevated electron temperatures may effect the recombination coefficients of the two species in different ways. It would be interesting to see if the rocket observations of elevated electron temperatures in the E region are sustained by subsequent analysis, since it is the electron temperature rather than the neutral or ion temperature which effects the recombination rate.

As the E-layer shape does not vary very much during an eclipse, and electron densities at different altitudes increase and decrease together, it follows that ionosonde measurements of critical frequency are perfectly satisfactory to describe E-layer eclipse behavior; however, it would be quite desirable to have a substantial network of stations; these need not necessarily be complete ionosondes, bistatic operation could well be used.

The similarity of behavior between the F1 and E layers has been demonstrated very well by Anastassiades, and is additional confirmation of the idea that the ionization in the F1-layer is primarily molecular, as is the E-layer ionization. Certainly, smaller values of recombination coefficient would be expected to occur in the F1 layer, due to the higher electron temperature, and generally speaking smaller values are found. It is by no means certain as to the optimum form of eclipse function to use for the F1 layer analysis; certainly there is some contribution from coronal lines in the wavelength region 175-200 Å. These would not be expected to fully eclipsed. On the other hand, a substantial part of the F1-layer ionizing radiation comes from chromospheric lines of helium and other elements. Probably, the exact eclipse circumstances are important. For an annular eclipse, for example, the chromosphere will certainly not be eclipsed; however, even for a total eclipse of short duration, it is possible that radiation from the chromosphere is still present in substantial amount, depending on the exact height in the solar atmosphere where it is produced. Effects occurring in a few seconds near totality, while they may not be important for the total density of the ionization, may nevertheless produce a substantial, almost instantanous effect on the electron temperature in the region where photoelectrons have the dominant heating effect.

Some very interesting observations of the F2-layer during solar eclipses have been presented at this meeting; however, no very coherent picture has emerged showing a consistent pattern of behavior of the F2 layer for all of the eclipses for which results are available. Certainly, dynamic effects are very important. For the lower part of the F2 layer, say 200-240 km, chemical effects are dominant, particularly at low latitudes, and a lag is seen in the minimum electron density, the lag increasing with altitude. At higher altitudes, however, and at medium latitudes, the behavior is much less consistent. In some eclipses, the lag continues to increase with altitude; in other eclipses the lag becomes zero or negative. These effects must surely result from changes in the thermal structure of the ionization, producing flow upwards or downwards from the topside. Thomas has shown the importance of diffusive equilibrium in the topside, particularly the effect of the transition between protons and atomic oxygen ions in determining the shape of the topside. Some very interesting observations were presented by

o
Matsaukas showing an increase of a total electron content during the solar eclipse
in May 1966; the explanation of this will be of great interest for theoreticians.
This single observation alone shows that chemical effects cannot be the whole
story in the eclipse behavior of the F layer, and that explanations must be
sought in transport processes of one kind or another. The idea of displacement
of atomic oxygen by protons is obviously an inviting one.

A further interesting conclusion from this conference was the possibility
of dynamic effects in the atmosphere at F region heights. Rycroft suggested
the excitation of atmospheric gravity waves by the eclipse; the identification
by several workers of disturbances which appear to travel from one station to
another in the early part of the eclipse, and the transitoria described by
Bibl are all dynamic phenomena which surely link together in some way which is
yet to be determined. For a future eclipse it would be very interesting to
concentrate particular attention on these effects.

While measurements from topside and bottomside sounders give a good picture
of the electron density morphology of the F region, the thermal structure is
so important, and plays such a dominant role in transport effects, that every
effort should be made to obtain eclipse results using Thomson-scatter sounders.
In addition, such sounders can measure directly the transport velocities in
the vertical direction, which can only be theorized otherwise.

Concerning the D region, the past several years have seen direct measurements
of eclipse effects in this region for the first time. Consequently, interpretation
of these effects is still in its infancy, and the relative roles of recombination
and attachment, for instance, are still far from clear. Particular emphasis will
be placed on the D region in the forthcoming eclipse program in United States in
March 1970, and rapid progress can be expected.

Eclipse effects occur rapidly, and therefore the experimental techniques
used need to have very good time resolution. In planning experiments, therefore,
it is necessary to consider the expected time scale of the anticipated processes
at the altitude studied, and then to choose the appropriate tool to accomplish
the measurements. Satellites, which would be ideal in their wide coverage, can
only measure at one or two points during the eclipse; similarly, Thomson-scatter
measurements have serious problems in that they require a finite time (several
minutes) to make the measurements, and the ionosphere changes substantially during
that time. Rockets again have the limitation that although they can measure
profiles at an single instant, they cannot make continuous observations. Therefore, there is necessarily a role for the older ground-based techniques which
can measure continuously throughout the eclipse. The necessity for collaboration
between the various techniques is one of the outstanding conclusions which has
emerged for me from this conference.

INDEX

Absorption
  A$_1$ type 188, 213
  differential 200, 203
  gyro-resonance 167
  ionospheric 187, 188
  normalized value 220
Active centers 256, 259, 276
Active regions 21
Aeronomics 46
Aerosols 69
Anomaly, equatorial 24
ARCAS 200

Bremsstrahlung 151, 167, 173
Brightness 158, 170, 171, 299

Calcium 128
Chapman 235
Charge 14, 88
Chemistry
  effects 9, 301
  hydrated protons 45
  ions 13, 57, 81
  nitric oxide 50
  particles 60
  photochemistry 70
Coefficient
  absorption 68
  a$_c$ 209
  a$_d$ 209
  correlation 230
  diffusion 49
  electron loss 10, 240
  ion-atom 10
  ionization loss 19
  neutralization 81

recombination 5, 6, 22, 83,
    210, 211, 261, 278

Collisions 11
Conjugate point 12
Constituents
  distribution 53
  major-minor 50
  neutral atmospheric 46
Continuity equation 5, 19, 20, 26
Contactivity 7, 24
Control days 11, 240, 267
Corona 2
  condensations 151, 152
  contour 261
  contribution 261, 300
  effect 261
  geocoronal radiation 241
  green line 6
  occultation 260
  prominences 260
  X-rays 219
Coulomb interactions 11
CRI 254

Dakar 264, 282
DASA 239
Dayglow 47
Diffusion
  downward 263
  eddy 49, 50
  fast 286
  plasma 24, 25
Disturbances 301
D-layer 301
  absorption effect 190

305